Memoirs of the American Mathematical Society

Number 433

Boris Youssin

Newton polyhedra without coordinates

Newton polyhedra of ideals

Published by the

AMERICAN MATHEMATICAL SOCIETY

Providence, Rhode Island, USA

September 1990 · Volume 87 · Number 433 (first of 3 numbers)

1980 *Mathematics Subject Classification* (1985 *Revision*).
Primary 13J05; Secondary, 14E15, 13B20.

Library of Congress Cataloging-in-Publication Data

Youssin, Boris, 1959-
Newton polyhedra without coordinates/Newton polyhedra of ideals/Boris Youssin.
p. cm. – (Memoirs of the American Mathematical Society, ISSN 0065-9266; no. 433)
Includes bibliographical references.
ISBN 0-8218-2495-3
1. Functions, Polyhedral. 2. Filters (Mathematics) 3. Rings (Algebra) 4. Ideals (Algebra)
I. Title. II. Series.
QA3.A57 no. 433
[QA343]
510 s–dc20 90-595
[515′.983] CIP

Subscriptions and orders for publications of the American Mathematical Society should be addressed to American Mathematical Society, Box 1571, Annex Station, Providence, RI 02901-1571. *All orders must be accompanied by payment.* Other correspondence should be addressed to Box 6248, Providence, RI 02940-6248.

SUBSCRIPTION INFORMATION. The 1990 subscription begins with Number 419 and consists of six mailings, each containing one or more numbers. Subscription prices for 1990 are $252 list, $202 institutional member. A late charge of 10% of the subscription price will be imposed on orders received from nonmembers after January 1 of the subscription year. Subscribers outside the United States and India must pay a postage surcharge of $25; subscribers in India must pay a postage surcharge of $43. Each number may be ordered separately; *please specify number* when ordering an individual number. For prices and titles of recently released numbers, see the New Publications sections of the NOTICES of the American Mathematical Society.

BACK NUMBER INFORMATION. For back issues see the AMS Catalogue of Publications.

MEMOIRS of the American Mathematical Society (ISSN 0065-9266) is published bimonthly (each volume consisting usually of more than one number) by the American Mathematical Society at 201 Charles Street, Providence, Rhode Island 02904-2213. Second Class postage paid at Providence, Rhode Island 02940-6248. Postmaster: Send address changes to Memoirs of the American Mathematical Society, American Mathematical Society, Box 6248, Providence, RI 02940-6248.

Printed in the United States of America.
The paper used in this book is acid-free and falls within the guidelines
established to ensure permanence and durability. ♾

10 9 8 7 6 5 4 3 2 1 95 94 93 92 91 90

Contents

NEWTON POLYHEDRA WITHOUT COORDINATES

1. Introduction 1

Chapter 1. Integrally closed filtrations

2. Definition and examples 6

3. The language of integrally closed subalgebras 8

4. Generators of a filtration 10

5. Join and suspension 11

6. Normal crossing rings and monomials 12

7. Special filtrations and the Main Theorem 15

Chapter 2. Contact and stably contact filtrations

8. Initial forms and transversality to the normal crossing divisor 20

9. Contact filtrations and their structure 22

10. Galois extensions 27

11. Stably contact filtrations 30

12. Special filtrations are stably contact 33

Chapter 3. The first derived filtration and its structure

13. First coweight and the first derived filtration $\mathrm{Dr}_1\,\Delta$ 37

14. Finitely Der-generated filtrations 40

15. Structure of $\mathrm{Dr}_1\,\Delta$— I. Properties of $\Delta^{(1)}$ 42

16. Structure of $\mathrm{Dr}_1\,\Delta$— II. The monomial 44

17. Structure of $\mathrm{Dr}_1\,\Delta$— III. $\Delta^{(1+)}$ 46

Chapter 4. Change of the subring

18. The result 49

19. The implications 52

20. Proof of the Main Theorem 7.11 55

Conclusion 21. Some unsolved problems 62

Appendices A1. Weights of a quasihomogeneous filtration are independent of the coordinate system 65

A2. Criterion for integral closedness 66

A3. The structure of suspension 68

A4. Multiplication and division by monomials: Proof of the structure theorem 69

A5. Suspension of a stably contact filtration is stably contact: Proof of (11.5) 71

References 74

1. Introduction 75
2. Standard bases and the main result 76
3. Differential operators and principal parts 78
4. Generalized Fitting ideals 80
5. Heuristics 82
6. Generic position 84
7. Fitting ideals and filtrations generated by standard bases 87
8. Normalized standard bases 91
9. Proof of the Main Theorem 2.7 93
Appendix. Sketch of another proof. 94
References 98

Abstracts

NEWTON POLYHEDRA WITHOUT COORDINATES

Given a function f in a complete regular ring of characteristic zero, we give a canonical construction of an ideal containing this function. This ideal in some local coordinate system is generated by the monomials in the coordinate functions whose exponents lie in some polyhedron Γ in the positive orthant. A similar result is proved for a system of functions $f_1, f_2, \ldots, f_n$.

For a function f , this construction gives us a coordinate-free construction of some Newton polyhedron related to f . This construction is also important for our future program of canonical resolution of singularities in characteristic zero.

Key words and phrases : Newton polyhedron, resolution of singularities, integrally closed filtration, quasi-homogeneous filtration, derivations in local rings.

NEWTON POLYHEDRA OF IDEALS

Given an ideal $\mathcal{A}$ in a complete regular local ring $\mathcal{O}$ of characteristic zero, we construct a canonical filtration in $\mathcal{O}$ associated with the ideal $\mathcal{A}$; this filtration consists of ideals which in some coordinate system are generated by the monomials in the coordinate functions whose exponents lie in some polyhedra in the positive orthant.

For an ideal $\mathcal{A}$, this construction gives us a coordinate-free definition of some Newton polyhedron related to $\mathcal{A}$. This construction is also important for our future program of canonical resolution of singularities in characteristic zero.

Key words and phrases : Newton polyhedron, resolution of singularities, contact filtration, standard base of an ideal.

NEWTON POLYHEDRA WITHOUT COORDINATES

Boris Youssin*

Acknowledgments

It is my pleasant duty to express my thanks to those with whom I discussed the questions of resolution of singularities over the past years: Mark Spivakovsky, A. N. Rudakov, A. Beilinson and his seminar, Jean Giraud, Heisuke Hironaka, David Kazhdan, and David Eisenbud. The impact of my discussions with A. G. Kouchnirenko in 1979–80 about the Newton polyhedra can be seen in Sections 1 and 21. Heisuke Hironaka, David Kazhdan, Mark Spivakovsky and the anonymous referee made very helpful suggestions on how to improve the manuscript. Finally, this work has been written while I was a graduate student in the warm Mathematics Department of Harvard University, where I was an Alfred P. Sloan Doctoral Dissertation Fellow. I am grateful to all these people and institutions.

1. Introduction.

Given a germ of functions f (algebraic, analytic, or just a formal power series) at some point of a smooth manifold, and a local coordinate system $x_1, x_2, \ldots, x_N$ at this point, we may define the Newton polyhedron Γ of f, $\Gamma \subset \mathbf{Q}^N$, as the convex hull of the exponents of all monomials involved in the power series expansion of f, as well as all monomials divisible by them. This polyhedron Γ contains important information about the function f; however, it depends on the choice of the coordinate system $x_1, x_2, \ldots, x_N$.

We may ask if there exists some "coordinate-free" definition of this polyhedron Γ. We may also define the ideal, which we denote by x^Γ, as the ideal generated by all the monomials in $x_1, x_2, \ldots, x_N$ whose exponents lie in Γ (then $f \in x^\Gamma$), and we may ask if this ideal may be defined independently of the coorinate system $x_1, x_2, \ldots, x_N$.

Of course, the answer to both of these questions is negative — the polyhedron Γ and the ideal x^Γ may change drastically as we change the coordinate system $x_1, x_2, \ldots, x_N$. Thus, we are forced to make some concessions, and the natural idea is to restrict the choice of the coordinate systems to some class of "good" coordinate systems determined by the function f. For example, we may consider only those coordinate systems that make the Newton polyhedron Γ minimal.

However, the examples show that already in the case of two variables there are some functions f for which there is no unique minimal Newton polygon Γ as we vary the coordinate system x_1, x_2.

Received by the editors May 18, 1988, and, in revised form, September 13, 1989.

* Alfred P. Sloan Doctoral Dissertation Fellow.

The idea that we propose here is as follows. We are no longer trying to give an invariant definition of the Newton polydron Γ of f in its entirety; instead, we want to get some partial information about it. This partial information is some *bigger* polyhedron Γ' containing Γ ; this polyhedron Γ' belongs to some special class of polyhedra that we define (see (7.5) below), and we prove the following (all these results assume the characteristic is zero and the ambient local ring is complete):

(1.1) *Given a coordinate system $x_1, x_2, \ldots, x_N$, there exists a unique minimal (in a certain sense that we define below in (7.8)) polyhedron Γ of our class, such that $f \in x^\Gamma$; this polyhedron Γ depends on f and the coordinate system $x_1, x_2, \ldots, x_N$.*

(1.2) *If we vary the coordinate system $x_1, x_2, \ldots, x_N$, then this polyhedron Γ achieves a unique minimum (again, we understand minimality in the same sense that we define in (7.8)).*

(1.3) *The ideals x^Γ are the same for all coordinate systems $x_1, x_2, \ldots, x_N$ for which this minimum is achieved.*

In other words, the statements (1.1)–(1.3) may be reformulated in the following way. We define some way of comparing the polyhedra Γ belonging to our class; then we may compare the ideals x^Γ for the different coordinate systems $x_1, x_2, \ldots, x_N$ just by comparing the respective polyhedra. In other words, we say that the ideal x^Γ is "smaller" than the ideal $(x')^{\Gamma'}$ if the polyhedron Γ is "smaller" than Γ' . (Here $x = (x_1, x_2, \ldots, x_N)$ and $x' = (x'_1, x'_2, \ldots, x'_N)$ are two coordinate systems and Γ, Γ' are two polyhedra of our class.) Then (1.1)–(1.3) amount to the following:

(1.4) *There is a unique minimal (in the sense of (7.8)) among the ideals x^Γ , where $x = (x_1, x_2, \ldots, x_N)$ is a coordinate system, Γ is a polyhedron of our class, and they satisfy the property $f \in x^\Gamma$.*

This allows us to associate in a canonical way a polyhedron Γ and an ideal x^Γ to any germ of functions f without reference to any particular coordinate system; on the contrary, our result specifies a class of coordinate systems in which the Newton polyhedron of f is minimal in our sense.

In the two-dimensional case this result takes the following form. Our special class of polyhedra (polygons, as the dimension is two) consists of polygons bounded by the two coordinate axes and a line crossing both of them; the corresponding ideals x^Γ are quasihomogeneous; they are generated by the monomials $x_1^{\alpha_1} x_2^{\alpha_2}$ where (α_1, α_2) satisfy the condition

$$\alpha_1 w_1 + \alpha_2 w_2 \geq 1$$

where $w_1, w_2 \geq 0$ are the weights. The ordering of these polygons is the lexicographical ordering of the weights (w_1, w_2) where we assume $w_1 \geq w_2$. The result (1.4) states in this case that there is a unique ideal containing a given function f , which is quasihomogeneous in some coordinate system and whose weights, taken in decreasing order, are lexicographically minimal. This quasihomogeneous ideal looks as shown on the picture.

Note that in the case shown in the left picture, the quasihomogeneous ideal corresponds to the first Puiseaux exponent of the solution of the equation $f = 0$.

A similar result holds in the N-dimensional case, too: given a germ of functions f , there exists a unique ideal containing it, which is quasihomogeneous in some coordinate system and whose weights, taken

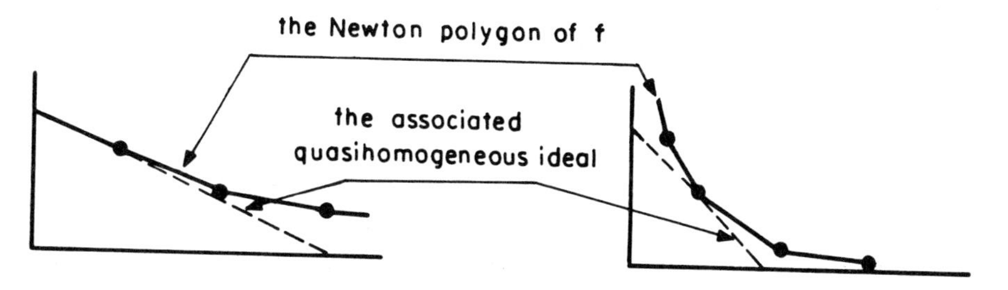

in the decreasing order, are lexicographyically minimal.* The general N-dimensional statement (1.4) that we prove in this work is a refinement of this statement; for the exact formulation, see (7.11) and, more generally, (20.10.1); these are the main results of this paper.

We also extend the result (1.4) to the case where we are given a finite number of functions $f_1, f_2, \ldots, f_n$ with positive rational weights $\nu_1, \nu_2, \ldots, \nu_n$ assigned to them; then we prove the following:

(1.5) *There is a unique minimal (in our sense) among the ideals* x^Γ , *where* $x = (x_1, x_2, \ldots, x_N)$ *is a coordinate system,* Γ *is a polyhedron of our class, and they satisfy the property* $f_i \in x^{\nu_i \Gamma}$, $i = 1, 2, \ldots, n$.

(Again, the exact formulation is in (7.11) and (20.10.1).)

This formulation prompts us to consider the family of ideals $\Delta(\nu) = x^{\nu\Gamma}$ for all rational ν ; they form a decreasing (as ν increases) filtration, and if Γ is convex — this is always the case if Γ belongs to our special class — then this filtration satisfies a natural integral closedness condition (see (2.1)).

Clearly, the statements (1.4) and (1.5) can be reformulated in terms of the existence and uniqueness of a filtration of the form $x^{\nu\Gamma}$ where Γ is a polyhedron which belongs to our class and $x = (x_1, x_2, \ldots, x_N)$ is a coordinate system (we call the filtrations of this kind "special filtrations"). Indeed, (1.4) states that there exists a unique special filtration Δ with the property $f \in \Delta(1)$, whose polyhedron Γ is minimal in our sense; similarly (1.5) states the same for the property $f_i \in \Delta(\nu_i)$, $i = 1, 2, \ldots, n$. Our way to prove these results is through the study of the properties of the special filtrations, and more generally, properties of the filtrations satisfying the integral closedness condition (2.1) ("integrally closed filtrations").

Of course, when we develop all this theory, we have an application in mind, and this application is the canonical resolution of singularities in characteristic zero. This means that we want to construct a resolution algorithm that would be functorial in the category of, say, étale morphisms.

[Hironaka 1] proved the possibility of resolution of singularities using induction by the embedding dimensions. He showed how to reduce any given resolution problem (e.g., a singular variety embedded in a smooth manifold) to some other resolution problems (of similar character) on certain smooth submanifolds of the ambient manifold. These submanifolds are not canonical, and Hironaka's resolution process a priori depends on their choice. Thus, a priori his resolution process is noncanonical.

In a subsequent paper we shall prove that in fact Hironaka's resolution process is independent of the choice of these submanifolds. The proof will be based on an explicit construction of the following objects:

* After this paper was written, I found out that this result was proved by [Milman] some ten years ago.

(a) a canonically defined smooth subvariety to be taken as a blowing-up center; and

(b) some measure of singularity that would strictly improve under such blowing-up (such a measure may be a string of numbers, and the improvement will then be understood lexicographically).

To construct these objects, we shall employ the results of this paper. Namely, given a singularity, we shall construct some canonically defined special filtration in the local ring of the ambient smooth manifold (here the word "special" has a technical meaning as above). In case our singularity is a hypersurface singularity with a local equation $f = 0$, this special filtration is the special filtration $x^{\nu\Gamma}$, where x^{Γ} is given by (1.4); in the general case, this special filtration is given by (1.5), where $f_1, f_2, \ldots, f_n$ is a certain system of equations of the singularity (the "normalized standard base" of [Hironaka 1], Definition 9, p. 248) and $\nu_1, \nu_2, \ldots, \nu_n$ are certain weights assigned to them. In another paper [Youssin 1] which appears at the end of this volume, we do this construction for the case when the ambient local ring is complete.

After this special filtration is constructed, the blowing-up center can be found as the intersection of certain prime components of this filtration, and the measure may be extracted from the numerical invariants of Γ. For the complete rings, this will be done in our next paper [Youssin 2].

In a subsequent paper we shall extend these results to the incomplete rings and show that our measure is globally semicontinuous. Moreover, we shall show that the blowing-up center mentioned above, is the "most singular" subset with respect to this measure (more precisely, the center is the subset where the measure is high enough). All this together will yield the canonical resolution of singularities in characteristic zero.

Note that there is another reason to look for a new proof — along these lines — of the resolution of singularities in characteristic zero. Indeed, the main difficulty of resolution of singularities in characteristic p is the impossibility to extend the Hironaka's idea of reduction to submanifolds into the positive characteristic (see the counterexample in [Narasimhan]). At the same time there are indications that other difficulties in characteristic p can be overcome (see [Moh]). This is why it is important to understand how to avoid reduction to submanifolds in a simpler case of characteristic zero.

Apparently, what we are trying to do here, is very close to what [Abhyankar] tried to do; unfortunately, his work has not been finished yet.

Idealistic exponents of [Hironaka 2] seem to be very close to (but still different from) our *finitely generated* integrally closed filtrations. As we do not need any results of [Hironaka 2], we do not study here the precise relationship between idealistic exponents and finitely generated integrally closed filtrations.

Our aim is, of course, very close to the aim of [Hironaka 3]; the relationship of the results of [Hironaka 3] with ours is as follows. [Hironaka 3] assumes that some part of the coordinate system is fixed (given in advance); and that part depends on the ideal (or the function f, in our notation) in question. At the expense of that he gets some more delicate information about the Newton polyhedra; on the contrary, we either do not assume any part of the coordinate system to be given in advance, and get information which is completely independent of any coordinate system, or we fix some part of the coordinate system (the normal crossing divisor, or the fixed variables of Section 6) without any reference to the function f in question, and get more delicate information on the Newton polyheddra, dependent on the "fixed variables".

Our ideas are the generalization of the ideas of [Hironaka 4] where the case of a surface embedded in a 3-manifold, is studied in the arbitrary characteristic. As [Moh] shows, these ideas cannot be extended directly into the case of positive characteristic and higher dimensions.

What differs our approach from the papers quoted above, is the emphasis on the ideals x^{Γ}. The results of [Hironaka 3] yield some Newton polyhedron related to the singularity; [Abhyankar] and [Hironaka 4] speak about expanding the equations in a certain way; we associate the ideals x^{Γ} to the singularity in a canonical way, and this is stronger than both other approaches.

For the reader's convenience, I tried to make this paper self-contained, requiring only the basic knowledge of algebraic geometry and commutative algebra; in particular, no previous knowledge of any works on resolution of singularities (not even [Hironaka 1]) is assumed. I should note, however, that many of the ideas and techniques applied here are taken from the previous works (which I tried to quote wherever appropriate).

The structure of the paper is as follows. In Chapter 1, we study the properties of integrally closed filtrations and some basic operations on them. In Chapter 2 we study the properties of two subclasses of the class of integrally closed filtrations — contact and stably contact filtrations — which are singled out by some good behavior under the derivations. These two classes of filtrations are smaller than the general class of integrally closed filtrations, but larger than the class of special filtrations.

The idea of the proof of our main result (1.5) is as follows. We first construct the minimal integrally closed filtration Δ such that $f_i \in \Delta(\nu_i)$, $i = 1, 2, \ldots, n$, then we construct the minimal contact and then the minimal stably contact filtration containing it. However, contact and stably contact filtrations have the property that they are essentially filtrations in a subring of a smaller dimension (more precisely, they are *suspensions* (see (5.7)) of filtrations in subrings). This allows us to use induction by dimension, and in Chapter 3 we define and study the propeties of the objects we need to make the induction step. Finally, in Chapter 4 we check that our procedure is independent of the choice of the subrings, and this completes the proof.

As we did not want to overload the main body of the paper, we moved some proofs and discussions of some side issues to the appendices.*

---

* After this work was written, I learned of the work [Bierstone, Milman] which independently developed very similar ideas in a slightly different context.

Chapter 1. Integrally closed filtrations

2. Definition and examples

All the rings that will appear in this paper will be commutative rings with identity. All the local rings in this paper will be understood to be Noetherian, but not necessarily of characteristic zero or even equicharacteristic, unless stated otherwise. By $\mathbf{Z}_+$ and $\mathbf{Q}_+$ we shall denote the sets of all nonnegative integers and rationals respectively.

Let $\mathcal{O}$ be any ring. We shall consider decreasing filtrations in $\mathcal{O}$ numbered by nonnegative rationals. A filtration Δ consists of the ideals $\Delta(\nu) \subset \mathcal{O}$ for each $\nu \in \mathbf{Q}_+$, such that if $\nu_1 < \nu_2$ then $\Delta(\nu_1) \supset \Delta(\nu_2)$. We shall also assume that for $\nu < 0$, $\Delta(\nu)$ is defined and equal to $\mathcal{O}$.

(2.1) Definition. *We shall say that a filtration Δ in a ring $\mathcal{O}$ is integrally closed if it is a decreasing filtration by ideals numbered by nonnegative rationals, and it has the following properties:*

(2.1.1) $\Delta(0) = \mathcal{O}$;

(2.1.2) $\Delta(\nu_1) \cdot \Delta(\nu_2) \subset \Delta(\nu_1 + \nu_2)$ *(i.e., the filtration agrees with the ring structure);*

(2.1.3) Integral closedness: if $g_1, g_2, \ldots, g_n$ are such that $g_i \in \Delta(i\nu)$ for some $\nu \in \mathbf{Q}_+$ and we can find $f \in \mathcal{O}$ such that

$$f^n + g_1 f^{n-1} + g_2 f^{n-2} + \ldots + g_n = 0$$

then $f \in \Delta(\nu)$.

(2.2) Example. Let $\Delta(\nu) = \mathcal{O}$ for each $\nu \in \mathbf{Q}_+$. Then Δ is an integrally closed filtration. We shall write $\Delta = (1)$.

(2.3) Example. Let $\mathcal{O}$ be a ring without nilpotents; let $\Delta(\nu) = (0)$ for $\nu \in \mathbf{Q}_+$, $\nu > 0$, and let $\Delta(0) = \mathcal{O}$. This is also an integrally closed filtration. We shall write $\Delta = (0)$.

We shall say that an integrally closed filtration Δ is *nontrivial* if $\Delta \neq (0)$ and $\Delta \neq (1)$.

(2.4) Example. Let $\mathcal{O}$ be a regular local ring with the maximal ideal $\mathcal{M}$, and let $\Delta(\nu) = \mathcal{M}^\nu$, where we understand $\mathcal{M}^\nu$ with ν rational by substituting the nearest greater integer for ν. Clearly, this is an integrally closed filtration ("the *maximal ideal filtration*").

(2.5) Example. Let $\mathcal{O} = \mathbf{k}[[x_1, x_2, \ldots, x_n]]$, where $\mathbf{k}$ is a field, and let Γ be a convex subset in the rational positive orthant $\mathbf{Q}_+^n$ which has the property

$$\Gamma + \mathbf{Q}_+^n = \Gamma \tag{2.5.1}$$

where the sets are added pointwise, and the points are added as vectors in $\mathbf{Q}_+^n$. (Another way to say this is that Γ contains $v + \mathbf{Q}_+^n$ together with each element $v \in \Gamma$.) Let $\Delta(\nu)$ be the ideal generated by those monomials

$$x^{\alpha} = x_1^{\alpha_1} x_2^{\alpha_2} \dots x_n^{\alpha_n}$$

whose exponents $\alpha = (\alpha_1, \alpha_2, \dots, \alpha_n)$ lie in $\nu \cdot \Gamma$. It is not hard to see that this filtration Δ is integrally closed (the integral closedness property (2.1.3) follows from the convexity of Γ). We shall call such a filtration *polyhedral*; clearly, the maximal ideal filtration is a particular case of a polyhedral filtration.

(2.6) Example. Another particular case of a polyhedral filtration is a *quasihomogeneous* filtration Δ in $\mathbf{k}[[x_1, x_2, \dots, x_n]]$. Its terms $\Delta(\nu)$ are generated by such monomials $x^{\alpha} = x_1^{\alpha_1} x_2^{\alpha_2} \dots x_n^{\alpha_n}$ whose exponents α satisfy $\sum \alpha_i w_i \geq \nu$, where $w_i \in \mathbf{Q}_+$ are the *weights* of the quasihomogeneous filtration. (The numbers $1/w_i$ will be referred to as *coweights*; note that we allow the weights to be zero and the coweights to be infinite.)

Now assume $\mathcal{O}$ is a regular local ring, and $x_1, x_2, \dots, x_n$ is a regular system of parameters in $\mathcal{O}$ (which we shall refer to as a coordinate system); then we may define a quasihomogeneous filtration in $\mathcal{O}$ in the same way as in $\mathbf{k}[[x_1, x_2, \dots, x_n]]$.

Clearly, the same filtration in $\mathcal{O}$ may be considered as quasihomogeneous in different coordinate systems. However, we could ask whether the weights of a quasihomogeneous filtration depend on the choice of the coordinate system.

(2.7) Proposition. *The weights $(w_1, w_2, \dots, w_n)$ of a quasihomogeneous filtration together with their multiplicities are defined uniquely up to permutation and do not depend upon the choice of the coordinate system.*

For the proof, see Appendix A1.

(2.8) Example. Let $\mathcal{O}$ be any ring and $f \in \mathcal{O}$ be a prime element (i.e., such that the principal ideal (f) is prime). Let $\Delta(\nu) = (f^{\nu})$, where in f^{ν} we substitute the nearest greater integer for the rational values of ν . Clearly, this filtration Δ is integrally closed; we shall say that Δ is *the filtration generated by* f . In Section 4 we shall generalized this definition for the case of f not being necessarily prime, and for the case of many generators with different weights.

(2.9) Remark-definition. *Let Δ be an integrally closed filtration, and let $\Delta'(\nu) = \Delta(k\nu)$, where k is a positive rational. Then Δ' is also an integrally closed filtration. We shall say that Δ' is a **rescaling** of the filtration Δ ; k is the **rescaling factor**. We shall write $\Delta' = \{\Delta'(\nu)\} = \{\Delta(k\nu)\}$; this notation understands the filtration Δ' as a collection of its terms $\Delta'(\nu) = \Delta(k\nu)$.*

Clearly, rescalings of both trivial filtrations (0) and (1) are the same trivial filtrations; rescaling of a polyhedral filtration with a polyhedron $\Gamma \subset \mathbf{Q}_+^n$ is a polyhedral filtration with the polyhedron $k\Gamma$; rescaling of a quasihomogeneous filtration changes its weights from $(w_1, w_2, \dots, w_n)$ to $(w_1/k, w_2/k, \dots, w_n/k)$, and the coweights — from $(1/w_1, 1/w_2, \dots, 1/w_n)$ to $(k/w_1, k/w_2, \dots, k/w_n)$. In particular, rescaling of the maximal ideal filtration is the quasihomogeneous filtration with the weights $(1/k, 1/k, \dots, 1/k)$. Rescaling of the filtration generated by f (Example 2.8) is not exactly the same kind of filtration, since f now has weight $1/k$; we shall call it the *filtration generated by* f *with the weight* $1/k$.

(2.10) Motivation for the definition of integrally closed filtration. The idea is to define some broad class of filtrations that would include *special filtrations* of (7.5).

Special filtrations will be defined as a subclass of polyhedral filtrations, containing quasihomogeneous filtrations. Thus, we have to consider quasihomogeneous filtrations with rational weights, and this explains why we need filtrations numbered by rationals and not just integers.

Special filtrations by their definition (see (7.5)) are constructed from each other using the operations of suspension (see (5.7)) and multiplication by fractional monomials (see (6.4)). In order to be able to multiply a filtration by a monomial, we have to be able to extract roots in the filtration, i.e., we need the property (2.1.3) to be satisfied only for the particular case of the equations of the form $f^n - g_n = 0$. However, taking suspension necessitates solving equations of the general form (2.1.3), as we can see from Appendix A3. Nevertheless, the difference between extracting roots and solving the equations of general form is not so great in our case, as explained in Appendix A2.

3. The language of integrally closed subalgebras

Here we show that integrally closed filtrations are in a natural one-to-one correspondence with integrally closed subalgebras of certain algebras.

Let $\mathcal{O}$ be any ring. By $\mathcal{O}[t^{\mathbf{Q}}]$ (resp. $\mathcal{O}[t^{\mathbf{Q}-}]$) we shall denote the $\mathbf{Q}$-algebra of all polynomials in the fractional powers t^α , $\alpha \in \mathbf{Q}$ (resp. $\alpha \in \mathbf{Q}$, $\alpha \leq 0$) of an independent variable t . We shall consider $\mathcal{O}[t^{\mathbf{Q}}]$ and $\mathcal{O}[t^{\mathbf{Q}-}]$ to be graded by the powers of t .

(3.1) Proposition. *Let Δ be an integrally closed filtration in $\mathcal{O}$. Then*

$$R_t(\Delta) = \sum_{\nu\in\mathbf{Q}_+} \Delta(\nu)t^\nu + \mathcal{O}[t^{\mathbf{Q}-}] = \sum_{\nu\in\mathbf{Q}} \Delta(\nu)t^\nu \subset \mathcal{O}[t^{\mathbf{Q}}] \tag{3.1.1}$$

is a graded $\mathcal{O}$-subalgebra of $\mathcal{O}[t^{\mathbf{Q}}]$; in addition, $R_t(\Delta)$ is integrally closed in $\mathcal{O}[t^{\mathbf{Q}}]$.

(3.2) Proof: Indeed, everything, except the integral closedness of $R_t(\Delta)$ in $\mathcal{O}[t^{\mathbf{Q}}]$, is obvious. As for the integral closedness, the following is clear: a homogeneous element of $\mathcal{O}[t^{\mathbf{Q}}]$ which is integral over $R_t(\Delta)$, lies in $R_t(\Delta)$ .Thus, it is enough to prove that, given a nonhomogeneous element integral over $R_t(\Delta)$, each of its homogeneous components is integral over $R_t(\Delta)$ too.

[Bourbaki], Ch. V, §1, Proposition 20, states this as a general statement about integral elements in $\mathbf{Z}$-graded rings. However, both $\mathcal{O}[t^{\mathbf{Q}}]$ and $R_t(\Delta)$, though $\mathbf{Q}$-graded, are unions of their $\mathbf{Z}$-graded subrings, namely,

$$\mathcal{O}[t^{\mathbf{Q}}] = \bigcup \mathcal{O}[t^{1/n}, t^{-1/n}]$$
$$R_t(\Delta) = \bigcup (R_t(\Delta) \cap \mathcal{O}[t^{1/n}, t^{-1/n}]) \ .$$

For any element $f \in \mathcal{O}[t^{\mathbf{Q}}]$ which is integral over $R_t(\Delta)$ and satisfies an equation

$$f^n + g_1 f^{m-1} + g_2 f^{m-2} + \ldots + g_m = 0$$

with $g_i \in R_t(\Delta)$, we may find such n that $f, g_1, g_2, \ldots, g_m \in \mathcal{O}[t^{1/n}, t^{-1/n}]$ and then the general statement of [Bourbaki] yields what we need. ■

(3.3) Proposition. *The correspondence (3.1.1) between the integrally closed filtrations Δ in $\mathcal{O}$ and the graded $\mathcal{O}$-algebras $R_t(\Delta)$, $\mathcal{O}[t^{\mathbf{Q}-}] \subset R_t(\Delta) \subset \mathcal{O}[t^{\mathbf{Q}}]$, which are integrally closed in $\mathcal{O}[t^{\mathbf{Q}}]$, is one-to-one.*

(Obvious.) ∎

As we shall see, sometimes the language of subalgebras is more convenient, and sometimes the language of filtrations is preferable.

(3.4) Remark. Filtration (0) corresponds to the subalgebra $\mathcal{O}[t^{\mathbf{Q}}-]$, and (1) to $\mathcal{O}[t^{\mathbf{Q}}]$. Filtration generated by f (2.8) corresponds to the subalgebra $\mathcal{O}[t^{\mathbf{Q}}-][ft] \subset \mathcal{O}[t^{\mathbf{Q}}]$; description of the subalgebras corresponding to the other examples of Section 2 is somewhat more complicated and it is left to the reader. We note only that the rescaling of a filtration (see (2.9)) corresponds to the substitution $t^{1/k}$ for t, where k is the rescaling factor.

(3.5) Definition. *Let Δ and Δ' be two integrally closed filtrations in a ring $\mathcal{O}$. We shall say Δ contains Δ' ($\Delta \supset \Delta'$) if, for each $\nu \in \mathbf{Q}_+$, $\Delta(\nu) \supset \Delta'(\nu)$.*

(3.6) Remark. Inclusions of filtrations clearly correspond to the inclusions of the corresponding subalgebras, i.e., $\Delta \supset \Delta'$ if and only if $R_t(\Delta) \supset R_t(\Delta')$.

(3.7) Remark. For any two integrally closed filtrations Δ and Δ' we may take their intersection $(\Delta \cap \Delta')(\nu) = \Delta(\nu) \cap \Delta'(\nu)$; it is easy to see that it is also an integrally closed filtration, and $R_t(\Delta \cap \Delta') = R_t(\Delta) \cap R_t(\Delta')$.

As an application of this approach, we shall prove the following

(3.8) Proposition. *Let Δ be a decreasing filtration in a ring $\mathcal{O}$ by ideals numbered by nonnegative rationals. Suppose it satisfies the properties (2.1.1) and (2.1.2). Consider a filtration Δ' in $\mathcal{O}$ defined by*

$$\Delta'(\nu) = \left\{ f \in \mathcal{O} \;\middle|\; \exists g_1, g_2, \ldots, g_n \in \mathcal{O},\ g_i \in \Delta(\nu i),\ \text{s.t. } f^n + g_1 f^{n-1} + \ldots + g_n = 0 \right\} \tag{3.8.1}$$

where $\nu \in \mathbf{Q}_+$. Then Δ' is an integrally closed filtration.

(3.9) Proof: Let

$$R_t(\Delta) = \sum_{\nu \in \mathbf{Q}_+} \Delta(\nu) t^\nu + \mathcal{O}[t^{\mathbf{Q}}-] \subset \mathcal{O}[t^{\mathbf{Q}}].$$

Then $R_t(\Delta)$ is a graded $\mathcal{O}$-subalgebra — this is equivalent to Δ being a decreasing filtration in $\mathcal{O}$ satisfying (2.1.1)–(2.1.2). Let

$$R_t(\Delta') = \sum_{\nu \in \mathbf{Q}_+} \Delta'(\nu) t^\nu + \mathcal{O}[t^{\mathbf{Q}}-] \subset \mathcal{O}[t^{\mathbf{Q}}].$$

Then (3.8.1) means that $R_t(\Delta')$ is the integral closure of $R_t(\Delta)$ in $\mathcal{O}[t^{\mathbf{Q}}]$ (cf. (3.2)). Thus, $R_t(\Delta')$ is integrally closed in $\mathcal{O}[t^{\mathbf{Q}}]$, so it corresponds to some integrally closed filtration. It is easy to see that this filtration coincides with Δ'. ∎

4. Generators of a filtration

(4.1) Proposition. *Let* $(f_\alpha)_{\alpha\in A}$ *be any family of elements of a ring* $\mathcal{O}$ *, and suppose we are given a number* $\nu_\alpha \in \mathbf{Q}_+$ *for each* f_α *. Then there is a minimal integrally closed filtration* Δ *with the property*

(4.1.1) $f_\alpha \in \Delta(\nu_\alpha)$ for all $\alpha \in A$.

Any other integrally closed filtration with the same property contains Δ *(in the sense of Definition 3.4).*

(4.2) Definition. *We shall say that this filtration* Δ *is generated by the family* $(f_\alpha)_{\alpha\in A}$ *with the weights* ν_α *,* $\alpha \in A$ *. If the family* $(f_\alpha)_{\alpha\in A}$ *is finite, then we shall say that* Δ *is finitely generated.*

(4.3) Proof of Proposition 4.1 : Clearly, condition (4.1.1) is equivalent to a condition for the corresponding subalgebras, namely, that

$$f_\alpha t^{\nu_\alpha} \in R_t(\Delta) \ .$$

Consider the intersection of all graded $\mathcal{O}$-subalgebras integrally closed in $\mathcal{O}[t^{\mathbf{Q}}]$ and containing $\mathcal{O}[t^{\mathbf{Q}-}]$ and $f_\alpha t^{\nu_\alpha}$ for all $\alpha \in A$. This intersection is also a graded $\mathcal{O}$-subalgebra integrally closed in $\mathcal{O}[t^{\mathbf{Q}}]$ and containing $\mathcal{O}[t^{\mathbf{Q}-}]$, so this intersection also corresponds to some integrally closed filtration. Clearly, this intersection contains $f_\alpha t^{\nu_\alpha}$ for all $\alpha \in A$, so the corresponding filtration Δ satisfies $f_\alpha \in \Delta(\nu_\alpha)$ for all $\alpha \in A$. It is easy to see that the filtration Δ has all the required properties. ■

(4.4) Remark. The subalgebra $R_t(\Delta)$ corresponding to the filtration Δ generated by the family $(f_\alpha)_{\alpha\in A}$ with the weights ν_α , $\alpha \in A$, is just the integral closure of $\mathcal{O}[t^{\mathbf{Q}-}][f_\alpha t^{\nu_\alpha}]_{\alpha\in A}$ in $\mathcal{O}[t^{\mathbf{Q}}]$.

(4.5) Remark. Filtration generated by a prime element f (Example 2.8) is a particular case of our construction when the family consists of only one element f which is assigned weight 1 .

(4.6) Corollary to Proposition 4.1. *Given a ring homomorphism* $f : \mathcal{O} \to \mathcal{O}'$ *and an integrally closed filtration* Δ *in* $\mathcal{O}$ *, there exists a minimal integrally closed filtration* $f_*\Delta$ *in* $\mathcal{O}'$ *satisfying* $(f_*\Delta)(\nu) \supset f(\Delta(\nu))$ *. Such a filtration* $f_*\Delta$ *is unique.* ■

An important case is when $\mathcal{O}'$ is the completion of $\mathcal{O}$ — this gives us the construction of the completion of an integrally closed filtration.

Another important case is when $\mathcal{O}'$ is a localization of $\mathcal{O}$ — this gives us a construction of a localization of an integrally closed filtration.

Similarly, we may consider $\mathcal{O}'$ to be a quotient ring of $\mathcal{O}$, e.g., $\mathcal{O}' = \mathcal{O}/\mathcal{A}$ where $\mathcal{A}$ is an ideal in $\mathcal{A}$. In this way we get a quotient filtration $\Delta \bmod \mathcal{A}$ in the ring $\mathcal{O}/\mathcal{A}$.

In case π is an inclusion, we shall sometimes identify $\pi_*\Delta$ with Δ (i.e., we shall write Δ instead of $\pi_*\Delta$, although it is an abuse of notation). A particular case when this abuse of notation does not cause any problem is

$$\pi : \mathcal{O} \hookrightarrow \mathcal{O}[[x_1, x_2, \ldots, x_m]]$$

where $x_1, x_2, \ldots, x_m$ are independent variables.

5. Join and suspension

(5.1) Corollary to Proposition 4.1. *Given two integrally closed filtrations Δ_1 and Δ_2, there exists a unique minimal integrally closed filtration Δ containing both of them.*

(5.2) Definition. *This filtration Δ is called the join of Δ_1 and Δ_2; notation: $\Delta = \Delta_1 * \Delta_2$.*

(5.3) Remark. Since $\Delta(\nu) \supset \Delta_1(\nu)$ and $\Delta(\nu) \supset \Delta_2(\nu)$, by multiplicativity we see that

$$\Delta(\nu) \supset \sum_{0 \leq \nu_1 \leq \nu} \Delta_1(\nu_1) \cdot \Delta_2(\nu - \nu_1)$$

(5.4) Remark. The definition of join immediately yields that it is an associative operation, i.e.

$$(\Delta_1 * \Delta_2) * \Delta_3 = \Delta_1 * (\Delta_2 * \Delta_3) .$$

(5.5) Example. Let Δ be a quasihomogeneous filtration in a regular local ring $\mathcal{O}$ with a coordinate system $x_1, x_2, \ldots, x_n$. Let $(w_1, w_2, \ldots, w_n)$ be its weights, and let $(c_1, c_2, \ldots, c_n)$ be its coweights, $c_i = 1/w_i$. Then

$$\Delta = \left\{(x_1^{c_1 \cdot \nu})\right\} * \left\{(x_2^{c_2 \cdot \nu})\right\} * \ldots * \left\{(x_n^{c_n \cdot \nu})\right\}$$

where $\left\{(x_i^{c_i \cdot \nu})\right\}$ is the rescaling of the filtration generated by x_i with the rescaling factor c_i (or, in other words, the filtration generated by x_i with the weight $w_i = 1/c_i$).

(5.6) Remark. If both Δ_1 and Δ_2 are polyhedral filtrations in a regular local ring $\mathcal{O}$ in the same coordinate system, then their join $\Delta = \Delta_1 * \Delta_2$ is clearly also a polyhedral filtration in the same coordinate system, and its polyhedron is the convex hull of the union of the polyhedra of Δ_1 and Δ_2.

(5.7) Definition. *Let Δ be an integrally closed filtration in a ring $\mathcal{O}$. Consider the following filtration in $\mathcal{O}[[x_1, x_2, \ldots, x_m]]$, where $x_1, x_2, \ldots, x_m$ are independent variables:*

$$\left\{(x_1, x_2, \ldots, x_m)^\nu\right\} * \Delta . \tag{5.7.1}$$

(Here we are identifying Δ with its image in $\mathcal{O}[[x_1, x_2, \ldots, x_m]]$, cf. (4.6).) We shall call this filtration the m-fold suspension of Δ.

(5.8) Remark. Since the join of two polyhedral filtrations is polyhedral, we see that the suspension of a polyhedral filtration is polyhedral too.

(5.9) Remark. Since suspension is a particular case of join, we may apply (5.3) which yields

$$\left[\left\{(x_1, x_2, \ldots, x_m)^\nu\right\} * \Delta\right](\nu_1) \supset \sum_{k=0}^{[\nu_1]+1} (x_1, x_2, \ldots, x_m)^k \cdot \Delta(\nu_1 - k) ,$$

where for any real ν we denote by $[\nu]$ the maximal integer not exceeding ν.

In general, (5.3) does not have to be an equality, as shown by Example 5.11 below. However, the following proposition shows that in case of suspension (5.3) is indeed an equality.

(5.10) Proposition. *The ν_1-th term of the suspension (5.7.1) is given by*

$$\left[\left\{(x_1, x_2, \ldots, x_m)^\nu\right\} * \Delta\right](\nu_1) = \sum_{k=0}^{[\nu_1]+1} (x_1, x_2, \ldots, x_m)^k \cdot \Delta(\nu_1 - k) .$$

For the proof, see Appendix A3.

(5.11) Example. (Suggested by Mark Spivakovsky.) We want to show that the right-hand side of (5.3) may be different from the left-hand side. Indeed,

$$\begin{aligned} \mathcal{O} &= \mathbf{k}[[x,y]] \\ \Delta_1(\nu) &= \{(y^\nu)\} \\ \Delta_2(\nu) &= \{(y^2+x^3)^\nu\} . \end{aligned}$$

Then $\Delta = \Delta_1 * \Delta_2$ is generated by y and y^2+x^3 with weight 1 assigned to both y and y^2+x^3 . Then it is easy to see that x^3 also has weight 1 (i.e., $x^3 \in \Delta(1)$); thus, x has weight 1/3 (i.e., $x \in \Delta(1/3)$). Thus, Δ is quasihomogeneous with the weights 1/3 and 1 assigned to x and y respectively.

Now let us look at both sides of (5.3) in this example, taking $\nu = 1/3$ .The left-hand side is just

$$\Delta(1/3) = (x,y) .$$

The right-hand side is equal to

$$\sum_{0 \le \nu_1 \le \nu} \Delta_1(\nu_1) \cdot \Delta_2(\nu - \nu_1) = (y, y^2 + x^3) = (y, x^3) .$$

We see that in this case (5.3) is not an equality.

6. Normal crossing rings and monomials

(6.1) Definition. *We shall say we are given a normal crossing ring if we are given a regular local ring $\mathcal{O}$ with a normal crossing divisor in* Spec $\mathcal{O}$ *; the components of this normal crossing divisor will be assumed to be ordered. In other words, we are given a local ring $\mathcal{O}$ with a (maybe empty) ordered set of principal ideals (u_i) , $i \in I$, which have the following property. Let $\mathcal{M}$ be the maximal ideal of $\mathcal{O}$; then the images in $\mathcal{M}/\mathcal{M}^2$ of these principal ideals (u_i) are linearly independent one-dimensional subspaces of $\mathcal{M}/\mathcal{M}^2$. The generators u_i , $i \in I$, of these ideals will be called fixed variables.*

By abuse of notation we shall use the same symbol for the normal crossing ring and for its underlying regular local ring.

We shall say that a normal crossing ring is *complete*, if the underlying regular local ring is complete.

(6.2) Remark. The fixed variables u_i, $i \in I$, of a normal crossing ring $\mathcal{O}$ are, of course, uniquely defined only modulo multiplication by invertible elements of $\mathcal{O}$.

These fixed variables u_i, $i \in I$, form a part of a coordinate system in $\mathcal{O}$. The opposite is also true: any ordered set of elements u_i, $i \in I$, which is a part of a coordinate system in a regular local ring $\mathcal{O}$, determines a structure of normal crossing ring on $\mathcal{O}$.

Note that in resolution of singularities the normal crossing divisor usually appears after some resolution work is done; it is formed by the exceptional submanifolds of the previous blowing-ups. In other words, the fixed variables are the local equations of the old exceptional divisors.

(6.3) Definition. *Let $\mathcal{O}$ be a normal crossing ring. Its formal power series extension by independent variables $x_1, x_2, \dots, x_n$ is a normal crossing ring whose underlying ring is $\mathcal{O}[[x_1, x_2, \dots, x_n]]$ and its fixed* *variables are that of $\mathcal{O}$ and some of $x_1, x_2, \dots, x_n$ — maybe none or all as well as anything in between. There could be different formal power series extensions depending on this choice; by abuse of notation we shall denote all of them by $\mathcal{O}[[x_1, x_2, \dots, x_n]]$.*

(6.4) Definition. *Let $\mathcal{O}$ be a normal crossing ring with fixed variables u_i , $i \in I$, and let Δ be an integrally closed filtration in it. Take $\alpha : I \to \mathbf{Q}_+$, and define an integrally closed filtration*

$$u^\alpha \cdot \Delta = \left(\prod_{i \in I} u_i^{\alpha(i)}\right) \cdot \Delta$$

as the minimal integally closed filtration having the property

$$(u^\alpha \cdot \Delta)(\nu) \supset \left(\prod_{i \in I} u_i^{\beta_i}\right) \cdot \Delta(\nu) \tag{6.4.1}$$

for each $\nu \in \mathbf{Q}_+$; here β_i is the minimal integer such that $\beta_i \geq \nu \cdot \alpha(i)$. In other words, we multiply each term $\Delta(\nu)$ by the minimal integer ("genuine") monomial

$$u^\beta = \prod_{i \in I} u_i^{\beta_i}$$

*in u_i, $i \in I$, which is divisible by the "fractional monomial" $u^{\nu\alpha}$, and take the integrally closed filtration generated by these ideals. We shall say that the filtration $u^\alpha \cdot \Delta$ is the result of **multiplication** of the filtration Δ by the **fractional monomial** u^α . (Of course, the fractional monomial u^α does not yet have any existence of its own!)*

(6.5) Definition. *In the notation of Definition 6.4 let $\Delta : u^\alpha$ be the minimal integrally closed filtration satisfying*

$$(\Delta : u^\alpha)(\nu) \supset \Delta(\nu) : \prod_{i \in I} u_i^{[\nu \cdot \alpha(i)]} . \tag{6.5.1}$$

*(Here $[\nu \cdot \alpha(i)]$ denotes, as usual, the maximal integer not exceeding $\nu \cdot \alpha(i)$.) We shall say that the filtration $\Delta : u^\alpha$ is the result of **division** of the filtration Δ by the fractional monomial u^α .*

(6.6) Example. Consider the filtration $u^\alpha \cdot (1)$. We can find $\nu \in \mathbf{Q}_+$, $\nu > 0$, such that $\nu\alpha(i) \in \mathbf{Z}_+$ for each i . Then $u^{\nu\alpha}$ is an integer monomial, i.e., it can be considered as an element of $\mathcal{O}$, and the filtration $u^\alpha \cdot (1)$ is generated by $u^{\nu\alpha}$ with the weight ν .

(6.7) Remark. The following is a straightforward corollary to the definitions:

(6.7.1) $u^\alpha \cdot (0) = (0) : u^\alpha = (0)$

(6.7.2) $(1) : u^\alpha = (1)$

(6.7.3) $u^\alpha \cdot u^\beta \cdot \Delta \subset u^{\alpha+\beta} \cdot \Delta$

(6.7.4) $(\Delta : u^\alpha) : u^\beta \subset \Delta : u^{\alpha+\beta}$

(6.7.5) $(u^\alpha \cdot \Delta) : u^\alpha \subset \Delta$

(6.7.6) $u^\alpha \cdot (\Delta : u^\alpha) \subset \Delta$.

As we shall see later (see Corollary 6.11), (6.7.3)–(6.7.5) are actually equalities, and (6.7.6) is an equality if $\Delta \subset u^\alpha \cdot (1)$.

(6.8) Definition. *Denote by* denom(α) *the minimal positive denominator of a rational number* α . *For a vector* α *with rational components or a function* α *with rational values* denom(α) *is the L.C.M. of the denominators of its components or values.*

(6.9) Proposition. *Let* Δ *be an integrally closed filtration in a normal crossing ring* $\mathcal{O}$ *with fixed variables* u_i , $i \in I$, *and let* u^α, $\alpha : I \to \mathbf{Q}_+$, *be a fractional monomial. Then*

$$(u^\alpha \cdot \Delta)(\nu) = \{f \in \mathcal{O} \mid f^n \in u^{n\nu\alpha} \cdot \Delta(n\nu),\ n = \text{denom}(\nu\alpha)\} \tag{6.9.1}$$

$$(\Delta : u^\alpha)(\nu) = \{f \in \mathcal{O} \mid u^{n\nu\alpha} \cdot f^n \in \Delta(n\nu),\ n = \text{denom}(\nu\alpha)\} \tag{6.9.2}$$

(For the proof, see Appendix A4.)

(6.10) Remark. Clearly, the right-hand sides of (6.9.1) and (6.9.2) are contained in the left-hand sides — this follows directly from the definitions, and it is the other inclusion which is nontrivial. It means that instead of multiplying or dividing $\Delta(\nu)$ for each $\nu \in \mathbf{Q}_+$ by an integer monomial that approximates the fractional monomial $u^{\nu\alpha}$, we may consider only those sufficiently large ν that $u^{\nu\alpha}$ is already an integer monomial, and after that use the integral closedness of our filtrations to extract roots.

(6.11) Corollary. *The inclusions (6.7.3)–(6.7.5) are actually always equalities, and (6.7.6) is an equality if and only if* $\Delta \subset u^\alpha \cdot (1)$.

(The proof is left to the reader.) ■

(6.12) Corollary. *Let* Δ_1 *and* Δ_2 *be two integrally closed filtrations in a normal crossing ring* $\mathcal{O}$, *and let* u^α *be a fractional monomial in* $\mathcal{O}$. *Then*

$$u^\alpha \cdot (\Delta_1 * \Delta_2) = (u^\alpha \cdot \Delta_1) * (u^\alpha \cdot \Delta_2)$$

(The *distribution property for join.*)

(6.13) Proof: Note that since $\Delta_1 * \Delta_2 \supset \Delta_i$, $i = 1, 2$,

$$u^\alpha \cdot (\Delta_1 * \Delta_2) \supset u^\alpha \cdot \Delta_1$$

for $i = 1, 2$, and thus,

$$u^\alpha \cdot (\Delta_1 * \Delta_2) \supset (u^\alpha \cdot \Delta_1) * (u^\alpha \cdot \Delta_2) . \tag{6.13.1}$$

To see the other inclusion, note that for $i = 1, 2$,

$$\Delta_i = (u^\alpha \cdot \Delta_i) : u^\alpha \subset \big[(u^\alpha \cdot \Delta_1) * (u^\alpha \cdot \Delta_2)\big] : u^\alpha .$$

(We used here the equality in (6.7.5) and the trivial inclusion $u^\alpha \cdot \Delta_i \subset (u^\alpha \cdot \Delta_1) * (u^\alpha \cdot \Delta_2)$.) Thus,

$$\Delta_1 * \Delta_2 \subset \big[(u^\alpha \cdot \Delta_1) * (u^\alpha \cdot \Delta_2)\big] : u^\alpha$$

and consequently

$$u^\alpha(\Delta_1 * \Delta_2) \subset u^\alpha \cdot \big[\big[(u^\alpha \cdot \Delta_1) * (u^\alpha \cdot \Delta_2)\big] : u^\alpha\big] \subset (u^\alpha \cdot \Delta_1) * (u^\alpha \cdot \Delta_2) . \tag{6.13.2}$$

Finally, (6.13.1) and (6.13.2) together prove (6.12). ■

(6.14) Corollary. *Suppose Δ is a polyhedral filtration in a coordinate system $x_1, x_2, \ldots, x_N$ such that each u_i, $i \in I$, is one of the coordinates $x_1, x_2, \ldots, x_N$. Then for any fractional monomial u^α the filtrations $u^\alpha \cdot \Delta$ and $\Delta : u^\alpha$ are also polyhedral. The polyhedron of $u^\alpha \cdot \Delta$ is obtained from the polyhedron of Δ by parallel transport by the vector $\widehat{\alpha} = (\widehat{\alpha}_1, \widehat{\alpha}_2, \ldots, \widehat{\alpha}_N) \in \mathbf{Q}_+^N$ which is defined by*

(6.14.1)
$$\widehat{\alpha}_j = \begin{cases} \alpha(i), & \text{if } x_j = u_i; \\ 0, & \text{if there is no such } i \in I. \end{cases}$$

(In other words, $x^{\widehat{\alpha}} = u^\alpha$, whatever this means.)

Finally, we get the polyhedron of $\Delta : u^\alpha$ if we take the polyhedron of Δ, apply the parallel transport by $-\widehat{\alpha}$, and then intersect the result with the positive orthant $\mathbf{Q}_+^N$.

(6.15) Proof: Indeed, all we have to do is to look at the right-hand sides of (6.9.1) and (6.9.2). ■

(6.16) Remark. Among other things, Corollary 6.14 states that the filtration $u^\alpha \cdot (1)$ is polyhedral; its polyhedron is $\widehat{\alpha} + \mathbf{Q}_+^N$.

(6.17) Corollary. *Δ is finitely generated if and only if $u^\alpha \cdot \Delta$ is finitely generated.*

(6.18) Proof: Indeed, if Δ is generated by $f_1, f_2, \ldots, f_n$ with the weights $\nu_1, \nu_2, \ldots, \nu_n$, then (6.9.1) shows that for some large N, $u^\alpha \cdot \Delta$ is generated by $u^{\nu_i \cdot N \cdot \alpha} \cdot f_i^N$, $i = 1, 2, \ldots, n$, with the weights $N\nu_1, N\nu_2, \ldots, N\nu_n$. This yields the "only if" part, and the "if" part is quite similar. ■

7. Special filtrations and the Main Theorem

In this section all the rings will be assumed to be complete equicharacteristic normal crossing rings. For a function $\mathrm{mon} : I \to \mathbf{Q}_+$ we shall denote

$$|\,\mathrm{mon}\,| = \sum_{i \in I} \mathrm{mon}(i).$$

(7.1) Definition. *The following filtrations in $\mathcal{O}$ will be called almost special:*

(7.1.1) trivial filtrations (0) and (1) and the maximal ideal filtration $\{\mathcal{M}^\nu\}$;

(7.1.2) $u^\alpha \cdot \Delta$, where Δ is any almost special filtration and u^α — a fractional monomial;

(7.1.3) rescaling $\{\Delta(k\nu)\}$ of any almost special filtration Δ;

(7.1.4) suspension $\{(x_1, x_2, \ldots, x_n)^\nu\} * \Delta'$, where Δ' is an almost special filtration in $\mathcal{O}'$, $\mathcal{O} = \mathcal{O}'[[x_1, x_2, \ldots, x_n]]$, $\Delta' \subset \{\mathcal{M}_{\mathcal{O}'}^\nu\}$, and $\mathcal{M}_{\mathcal{O}'}$ is the maximal ideal of $\mathcal{O}'$.

(Whenever we write an equality like $\mathcal{O} = \mathcal{O}'[[x_1, x_2, \ldots, x_n]]$, we shall understand it as an equality of normal crossing rings — see (6.3).)

(7.2) Proposition. *Let Δ be an almost special filtration in $\mathcal{O}$. Then we can find such coordinate system $x_1, x_2, \ldots, x_N$ in $\mathcal{O}$, $\mathcal{O} = \mathbf{k}[[x_1, x_2, \ldots, x_N]]$ (here $\mathbf{k}$ is a field and this equality is understood as an equality of normal crossing rings), such integers $0 \le n_1 < n_2 < \ldots < n_k < N$, such fractional monomials $u^{\mathrm{mon}_1}, u^{\mathrm{mon}_2}, u^{\mathrm{mon}_3}, \ldots, u^{\mathrm{mon}_k}$, and such rational numbers $c_1, c_2, c_3, \ldots, c_{k+1}$ (c_{k+1} only may also be equal to ∞) that*

(7.2.1) each u^{mon_i}, $1 \le i \le k$ is a fractional monomial in $\mathbf{k}[[x_{n_i+1}, x_{n_i+2}, \ldots, x_N]]$,

(7.2.2) *each of* $c_1, c_2, \ldots, c_k$ *is a positive rational, and* c_{k+1} *may be either a nonnegative rational or* ∞ ,

(7.2.3) $\Delta = \{(x_1, x_2, \ldots, x_{n_1})^{c_1 \nu}\} * [u^{\mathrm{mon}_1} \cdot \{(x_{n_1+1}, x_{n_1+2}, \ldots, x_{n_2})^{c_2 \nu}\}]$
$* [u^{\mathrm{mon}_1 + \mathrm{mon}_2} \cdot \{(x_{n_2+1}, x_{n_2+2}, \ldots, x_{n_3})^{c_3 \nu}\}] * \ldots$
$* [u^{\mathrm{mon}_1 + \mathrm{mon}_2 + \ldots + \mathrm{mon}_k} \cdot \{(x_{n_k+1}, x_{n_k+2}, \ldots, x_N)^{c_{k+1} \cdot \nu}\}]$,

(7.2.4) $|\mathrm{mon}_i| + c_{i+1} \geq c_i$.

(7.3) Remark. Cases $c_{k+1} = 0$ and $c_{k+1} = \infty$ in (7.2.3) are understood in the following natural way:

$$\{\mathcal{M}^{0 \cdot \nu}\} = (1)$$
$$\{\mathcal{M}^{\infty \cdot \nu}\} = (0)$$

Note that in case $c_{k+1} = \infty$ the filtration Δ does not change if we change mon_k ; to avoid this ambiguity, we shall assume that in this case $\mathrm{mon}_k = \infty$, i.e., $\mathrm{mon}_k(i) = \infty$ for each $i \in I$ such that u_i is one of $x_{n_k+1}, x_{n_k+2}, \ldots, x_N$.

(7.4) Proof of Proposition 7.2: Indeed, by Definition 7.1 we should start with either of (0) , (1) , $\{\mathcal{M}^\nu\}$, and then apply rescalings, multiplication by monomials, and suspensions, and after a finite number of steps we shall get Δ . Clearly, there is no point in applying two rescalings or two suspensions or two multiplications by monomials in a row; thus, we may assume these operations are applied alternately.

Note that rescalings and multiplications by monomials almost commute; more precisely, the composition of a rescaling and a multiplication by a monomial coincides with the composition of a multiplication by another monomial and the same rescaling. Thus, any sequence of rescalings and multiplications by monomials (in any order) is equivalent to just one rescaling (first) and one multiplication by a monomial (second).

This shows that Δ can be obtained from one of (0) , (1) , $\{\mathcal{M}^\nu\}$ by a following sequence of operations: rescaling, multiplication by a monomial, suspension, rescaling, multiplication by a monomial, suspension, etc. (Note that we allow some rescalings to be trivial — with rescaling factor 1 — and some multiplications by monomials to be also trivial — with monomial $u^\alpha = 1$, i.e. $\alpha = 0$. However, we do require some nontrivial operation between every two suspensions, and all the suspensions are nontrivial.)

The sequence may end by either of the three kinds of operations. For the sake of uniformity, we shall assume that this sequence always ends by a rescaling; we achieve this by putting at the end, if necessary, a trivial rescaling and/or a fictitious "0-fold suspension". (Of course, all other suspensions in the sequence are still nontrivial.)

We can also make the beginning of this sequence uniform by incorporating the three starting cases (0) , (1) , $\{\mathcal{M}^\nu\}$ into one by allowing the result of the first rescaling to be $\{\mathcal{M}^{c\nu}\}$. where $c \in \mathbf{Q}_+ \cup \{\infty\}$.

Now it is easy to see that this sequence of operations gives us Δ exactly in the form (7.2.3), where c_i and u^{mon_i} satisfy (7.2.1) and (7.2.2).

To complete the proof of Proposition 7.2, we need to show also (7.2.4). Indeed, (7.1.4) allows us to take suspensions of only such filtrations Δ' that $\Delta' \subset \{\mathcal{M}^\nu\}$, and this immediately yields (7.2.4). ■

(7.5) Definition. *We shall say that an almost special filtration given by (7.2.3) is special if the following conditions are satisfied:*

(7.5.1) $n_1 > 0$;

(7.5.2) If $c_i > c_{i+1}$ *for any* i , $1 \le i \le k$, *then*

$$u^{\frac{c_{i+1}}{c_i - c_{i+1}} \cdot \mathrm{mon}_i} \cdot (1) \subset \{(x_{n_i+1}, x_{n_i+2}, \ldots, x_{n_{i+1}})^{c_{i+1} \cdot \nu}\}$$
$$* \left[u^{\mathrm{mon}_{i+1}} \cdot \{x_{n_{i+1}+1}, x_{n_{i+1}+2}, \ldots, x_{n_{i+2}})^{c_{i+2} \cdot \nu}\}\right]$$
$$* \left[u^{\mathrm{mon}_{i+1} + \mathrm{mon}_{i+2}} \cdot \{(x_{n_{i+2}+1}, x_{n_{i+2}+2}, \ldots, x_{n_{i+3}})^{c_{i+3} \cdot \nu}\}\right]$$
$$* \cdots$$
$$* \left[u^{\mathrm{mon}_{i+1} + \mathrm{mon}_{i+2} + \ldots + \mathrm{mon}_k} \cdot \{(x_{n_k+1}, x_{n_k+2}, \ldots, x_N)^{c_{k+1} \cdot \nu}\}\right]$$

(7.5.3) $|\mathrm{mon}_i| + c_{i+1} > c_i$.

(7.6) Remark. Similarly to the way we defined the almost special filtrations in (7.1), we may define the special filtrations in the following inductive way:

(i) filtrations (0) , (1) , and $\{\mathcal{M}^\nu\}$ are special;

(ii) rescaling of a special filtration is special;

(iii) suspension of a filtration Δ is special, provided $\Delta \subset \{\mathcal{M}^\nu\}$ and $\Delta = u^\alpha \cdot \Delta'$, where Δ' is a special filtration satisfying the following conditions. Let c be the maximal number such that $\Delta' \subset \{\mathcal{M}^{c\nu}\}$ (this means that Δ' is given by the formula (7.2.3) with $c_1 = c$). Then

(7.6.1) $|\alpha| + c > 1$;

(7.6.2) either $c \ge 1$ or

$$u^{\frac{c}{1-c} \cdot \alpha} \cdot (1) \subset \Delta' .$$

(Note that the filtration Δ in (iii) is not special.)

The reader who is familiar with the original proof of Hironaka's, will notice that the last condition corresponds to the construction for the case $\nu < b$ on p. 323 of [Hironaka 1]. Strangely, the property (7.6.2) also plays the key role in our Main Theorem 7.11 (cf. also (12.3), (17.7)). Example 21.6 shows that the Main Theorem 7.11 is not true if this condition is dropped (see also the discussion in (21.6)–(21.7)).

(7.7) Remark. Clearly, almost special filtrations are polyhedral, since (0) , (1) , and $\{\mathcal{M}^\nu\}$ are polyhedral filtrations and the operations of rescaling, multiplication by a monomial and suspension preserve the class of polyhedral filtrations. As special filtrations are a particular case of almost special filtrations, they are polyhedral too.

The special filtrations that we defined here are the special filtrations that we mentioned in Section 1, and the "special class of polyhedra" that we mentioned there, is just the class of polyhedra that correspond to the special filtrations defined here.

In case of an empty normal crossing divisor (i.e., no fixed variables at all) the classes of special, almost special, and quasihomogeneous filtrations coincide. The same is true in the case when there are fixed variables, but all the monomials are trivial ($u^{\mathrm{mon}_i} = 1$ and $\mathrm{mon}_i = 0$ for all i).

In Appendix A1 we prove that the weights of a quasihomogeneous filtration are intrinsically defined by the filtration. In (19.13) we generalize this statement to show that all the numerical data $n_1, n_2, \ldots, n_k$, $c_1, c_2, \ldots, c_{k+1}$, $\mathrm{mon}_1, \mathrm{mon}_2, \ldots, \mathrm{mon}_k$ of a special filtration are also intrinsically defined by the filtration.

(7.8) Definition. *Let* Δ *be a special filtration given in the form (7.2.3), and in the notation of (7.2.3) let* $m_1 = n_1$, $m_i = n_i - n_{i-1}$ *for* $2 \le i \le k$, $m_{k+1} = N - n_k$.

We define the first characteristic string of Δ *to be the following string of numbers:*

$$\operatorname{char}_1 \Delta = (c_1, -m_1, |\operatorname{mon}_1|, c_2, -m_2, |\operatorname{mon}_2|, c_3, -m_3, |\operatorname{mon}_3|, \ldots) \tag{7.8.1}$$

where the last element of the string is either c_{k+1} *, if it is* 0 *or* ∞ *, or* $-m_{k+1}$ *, if* $0 < c_{k+1} < \infty$ *.*

Recall that in case $c_{k+1} = \infty$ we have assumed $\operatorname{mon}_k = \infty$ (cf.(7.3)); thus, in this case $|\operatorname{mon}_k| = \infty$. Note that all the other elements of the first characteristic string are always finite.

We shall assume the first characteristic strings are *ordered* lexicographically, the elements on the left being the most significant.

The ordering of the "special class of polyhedra" (or, equivalently, of the class of special filtrations) which was mentioned in Section 1, is the ordering of their first characteristic strings; a polyhedron (or a special filtration) is "smaller" if its first characteristsic string is lexicographically larger. (Note that an inequality between the first characteristic strings of two special filtrations does not necessarily imply the inclusion of the filtrations; the converse — that the inclusion of special filtrations implies the inequality between their first characteristic strings — is actually one of the statements of the Main Theorem 7.11.)

(7.9) Remark. Consider the first characteristic strings of any two special filtrations in the same ring $\mathcal{O}$. Then neither of them can be a continuation of the other one to the right; this is because the first characteristic string terminates either at $c_{k+1} = 0$, or at $c_{k+1} = \infty$, or at

$$\sum_{\ell=1}^{k+1} m_\ell = N = \dim \mathcal{O} ,$$

whichever of these conditions comes first. Thus, for any two special filtrations Δ_1, Δ_2 in $\mathcal{O}$, we can either say which of $\operatorname{char}_1 \Delta_1$, $\operatorname{char}_1 \Delta_2$ is lexicographically larger, or $\operatorname{char}_1 \Delta_1 = \operatorname{char}_1 \Delta_2$. In other words, char_1 is a function on special filtrations with the values in a totally ordered set.

Note also that a priori the first characteristic string $\operatorname{char}_1 \Delta$ depends on the choice of the presentation of Δ in the form (7.2.3), and the same filtration Δ may be presented in the form (7.2.3) in different coordinate systems. Later in (19.13) we shall establish that $\operatorname{char}_1 \Delta$ is an intrinsic invariant of Δ; before that we should bear in mind that $\operatorname{char}_1 \Delta$ is actually a function of both Δ and its presentation in the form (7.2.3).

(7.10) Remark. In case of quasihomogeneous filtrations (i.e., when $\operatorname{mon}_i = 0$ for all i) the ordering of the first characteristic strings is the same as the lexicographical ordering of coweight sequences $(1/w_1, 1/w_2, \ldots, 1/w_N)$, assuming the coweights are ordered in increasing order $1/w_1 \leq 1/w_2 \leq \ldots \leq 1/w_N$ and the dimension N of the ambient ring is fixed.

(7.11) Main Theorem. *Given a finitely generated integrally closed filtration* Δ *in a complete normal crossing ring* $\mathcal{O}$ *of characteristic zero, there exists a unique special filtration* Δ^* *in* $\mathcal{O}$ *with the following property: it contains* Δ *and its first characteristic string is the largest among the first characteristic strings of all special filtrations containing* Δ *. (We shall say that* Δ^* *is the* **special filtration associated to** Δ *.) In addition, if* Δ *is already a special filtration, then* $\Delta^* = \Delta$ *.*

In other words, given any elements $f_1, f_2, \ldots, f_n \in \mathcal{O}$, and any weights $\nu_1, \nu_2, \ldots, \nu_n$, there exists a unique special filtration Δ^* in $\mathcal{O}$ such that $f_i \in \Delta^*(\nu_i)$ and such that $\operatorname{char}_1 \Delta^*$ is maximal.

The last statement of the Main Theorem 7.11 — which concerns the case of Δ being a special filtration — means that an inclusion of special filtrations, say, $\Delta_1 \subset \Delta_2$, implies the inequality of their first characteristic strings: $\mathrm{char}_1 \Delta_1 \geq \mathrm{char}_1 \Delta_2$, and if, in addition, $\Delta_1 \neq \Delta_2$, then $\mathrm{char}_1 \Delta_1 > \mathrm{char}_1 \Delta_2$. (Cf. the discussion in (7.8).)

(7.12) Remark. We say "the first characteristic string" because there is another ("the second", of course) characteristic string that strictly decreases under the blowings-up with appropriately chosen centers. This second string is the "measure of singularity" mentioned in Section 1; it is introduced in [Youssin 2].

Chapter 2. Contact and stably contact filtrations

In this chapter we introduce two properties of integrally closed filtrations, being contact and being stably contact, and show that the special filtrations of Section 7 (on certain conditions) have these properties.

Given an integrally closed filtration Δ , we construct the minimal contact filtration and the minimal stably contact filtration containing it; in Chapter 3 we shall apply this machinery to construct the first derived filtration $\mathrm{Dr}_1\,\Delta$ which is the first step towards the construction of the minimal special filtration $\Delta^* \supset \Delta$ of the Main Theorem 7.11.

8. Initial forms and transversality to the normal crossing divisor

Let V be a finite-dimensional vector space over a field $\mathbf{k}$. Let

$$\mathbf{k}[V] = \bigoplus_{n=0}^{\infty} S^n V ,$$

where $S^n V$ is the n-th symmetric power of V . The natural multiplication

$$S^n V \otimes S^m V \to S^{n+m} V$$

defines the structure of a graded $\mathbf{k}$-algebra on $\mathbf{k}[V]$; if $x_1, x_2, \ldots, x_N$ is a base of V , the $\mathbf{k}[V] = \mathbf{k}[x_1, x_2, \ldots, x_N]$; otherwise we may consider $\mathbf{k}[V]$ as the algebra of all polynomial functions on the dual space V^* . The following lemma is well-known.

(8.1) Lemma. *Let $\mathcal{A} \subset \mathbf{k}[V]$ be a homogeneous ideal (respectively, a homogeneous subring). Then there is a minimal subspace $W \subset V$ such that $\mathcal{A} = (\mathcal{A} \cap \mathbf{k}[W]) \cdot \mathbf{k}[V]$ (respectively, $\mathcal{A} \subset \mathbf{k}[W]$), and any $W' \subset V$ such that $\mathcal{A} = (\mathcal{A} \cap \mathbf{k}[W']) \cdot \mathbf{k}[V]$ (respectively, $\mathcal{A} \subset k[W']$) satisfies $W' \supset W$.*

For the case when $\mathcal{A}$ is an ideal, this lemma is due to [Hironaka 1], Lemma 10, p. 221. Geometrically, the ideal $\mathcal{A}$ defines a cone C in V^* , and the subspace $W^\perp$ (the orthogonal complement to W in V^*) is the maximal subspace in V^* such that $C + W^\perp = C$ (addition here is understood as the pointwise addition in V^*). In case $\mathcal{A}$ is the ideal of the tangent cone to a singular algebraic variety at a point, [Hironaka 4], p. 100 calls $W^\perp$ the *strict tangent space* at this point. This subspace $W^\perp$ is related to the concept of *ridge of a cone* due to [Giraud 2], I.5.2, [Giraud 3], 1.5.

(8.2) Proof: Consider the action of V^* on $\mathbf{k}[V]$ as derivations. If $\operatorname{char}\mathbf{k} = 0$, then clearly $\mathcal{A} = (\mathcal{A} \cap \mathbf{k}[W]) \cdot \mathbf{k}[V]$ (respectively, $\mathcal{A} \subset \mathbf{k}[W]$) if and only if $W^\perp \cdot \mathcal{A} \subset \mathcal{A}$ (respectively, $W^\perp \cdot \mathcal{A} = 0$). Clearly, there is a maximal subspace W_1 in V^* such that $W_1 \cdot \mathcal{A} \subset \mathcal{A}$ (respectively, $W_1 \cdot \mathcal{A} = 0$); then $W = W_1^\perp$ is the minimal subspace in V such that $\mathcal{A} = (\mathcal{A} \cap \mathbf{k}[W]) \cdot \mathbf{k}[V]$ (respectively, $\mathcal{A} \subset \mathbf{k}[W]$).

In case $\operatorname{char}\mathbf{k} = p > 0$ we should also consider the actions of V^* in $\mathbf{k}[V]$ by Hasse derivations; if $x_1, x_2, \ldots, x_n$ is a basis of V , then each $\partial/\partial x_i$ (which is an element of V^*) acts on $\mathbf{k}[V]$ by the Hasse derivation

$$\frac{1}{p^n!}\left(\frac{\partial}{\partial x_i}\right)^{p^n} : x_i^\ell \mapsto \binom{\ell}{p^n} x_i^{\ell-p^n}$$

for each n. We should take the maximal $W_1 \subset V^*$ such that it preserves (respectively, vanishes on) $\mathcal{A}$ under all these actions, then again $W = W_1^\perp$. ■

(8.3) Remark. Let $f \in \mathbf{k}[V]$ be a homogeneous element of degree n. If $\operatorname{char}\mathbf{k} = 0$, then we may obtain the minimal subspace $W \subset V$ such that $f \in \mathbf{k}[W]$, by differentiating f, namely, $(n-1)$ times:

$$W = (V^*)^{n-1} \cdot f$$

For a subalgebra, we should do the same for each of its homogeneous generators and take the sum of the resulting subspaces in V; for an ideal, we have to do the same for all the elements of its minimal ("standard") base introduced by [Hironaka 1], pp. 205–209 (see also in this volume [Youssin 1], (2.4)).

The case of $\operatorname{char}\mathbf{k} > 0$ is more complicated — it is not even enough to use Hasse derivations to get the generalization of this approach.

Now assume we are also given some one-dimensional subspaces

$$L_1, L_2, \ldots, L_k \subset V$$

which are transverse to each other, i.e., $\dim(L_1 + L_2 + \ldots + L_k) = k$.

(8.4) Definition. *We shall say that a linear subspace $W \subset V$ is **weakly transverse** to $L_1, L_2, \ldots, L_k$ if W is transverse to all L_i, $1 \le i \le k$, that are not contained in W, i.e.,*

$$\dim(W + L_1 + L_2 + \ldots + L_k) = \dim W + \#\{\, L_i \mid L_i \not\subset W \,\}.$$

(8.5) Remark. Clearly, W is weakly transverse to $L_1, L_2, \ldots, L_k$ if and only if there is a coordinate system in V such that $L_1, L_2, \ldots, L_k$ are coordinate axes and W is a coordinate plane with respect to this coordinate system.

(8.6) Lemma. *For any linear subspace $W \subset V$ there is a minimal linear subspace $\widetilde{W} \supset W$ which is weakly transverse to $L_1, L_2, \ldots, L_k$.*

(8.7) Proof: Consider $W_1 = W \cap (L_1 + L_2 + \ldots + L_k)$; clearly, W is weakly transverse to $L_1, L_2, \ldots, L_k$ if and only if $W_1 = \sum_{i\in J} L_i$ for some $J \subset \{1, 2, \ldots, k\}$. For each $i = 1, 2, \ldots, k$ there is a projection $W_1 \to L_i$ induced by the projection $L_1 + L_2 + \ldots + L_k \to L_i$. Let

$$J = \{\, i \mid 1 \le i \le k, \text{ the projection } W_1 \to L_i \text{ is nonzero } \}.$$

Let $W_2 = \sum_{i\in J} L_i$; then W_2 is the minimal subspace in $L_1 + L_2 + \ldots + L_k$ of the form $\sum L_i$ which contains W_1. It is easy to see that $\widetilde{W} = W + W_2$ is the minimal subspace of V which is weakly transverse to $L_1, L_2, \ldots, L_k$ and contains W. ■

(8.8) Definition. *Let V, $L_1, L_2, \ldots, L_k$, W, $\widetilde{W}$ be as above. We shall say that W is **intranverse** to L_i if $L_i \subset \widetilde{W}$.*

(8.9) Remark. If $J \subset \{1, 2, \ldots, k\}$ is the set of such i that W is intransverse to L_i, then $\widetilde{W} = W + \sum_{i\in J} L_i$ is the minimal subspace of V which is weakly transverse to $L_1, L_2, \ldots, L_k$.

(8.10) Definition. *Let V, $L_1, L_2, \ldots, L_k$ be as above, let $\mathcal{A} \subset \mathbf{k}[V]$ be a homogeneous ideal (respectively, a homogeneous subring), and let W be the minimal subspace of V constructed in Lemma 8.1. We shall say that $\mathcal{A}$ is intransverse to L_i if W is intransverse to L_i.*

(8.11). Now let $\mathcal{O}$ be a normal crossing ring with the maximal ideal $\mathcal{M}$, $\mathcal{O}/\mathcal{M} = \mathbf{k}$, and let $V = \mathcal{M}/\mathcal{M}^2$. Then $\operatorname{gr}\mathcal{O} = \mathbf{k}[V]$ and the fixed variables u_i, $i \in I$, define one-dimensional subspaces $L_i = \left[(u_i) \bmod \mathcal{M}^2\right] \subset V$.

Let Δ be an integrally closed filtration in $\mathcal{O}$ satisfying $\Delta \subset \{\mathcal{M}^\nu\}$. Then

(8.11.1) $$\bigoplus_{n=0}^{\infty}\left[\Delta(n) \bmod \mathcal{M}^{n+1}\right] \subset \operatorname{gr}\mathcal{O} = \mathbf{k}[V]$$

is a homogeneous subring in $\mathbf{k}[V]$. (In fact, this is true for any filtration Δ satisfying $\Delta \subset \{\mathcal{M}^\nu\}$ and the multiplicativity conditions (2.1.1) and (2.1.2), while integral closedness (2.1.3) is not important here.)

(8.12) Definition. *Let Δ be as in (8.11). We shall say that the subring*

$$\operatorname{Init}(\Delta) = \bigoplus_{n=0}^{\infty}\operatorname{Init}_n(\Delta) = \bigoplus_{n=0}^{\infty}\left[\Delta(n) \bmod \mathcal{M}^{n+1}\right] \subset \mathbf{k}[V]$$

is the initial form ring of Δ.

(8.13) Definition. *We shall say that the filtration Δ, $\Delta \subset \{\mathcal{M}^\nu\}$, is intransverse to a fixed variable u_i, $i \in I$, if $\operatorname{Init}(\Delta)$ is intransverse to L_i (see Definition 8.10).*

We shall denote by $I_{\mathrm{tr}}(\Delta)$ the set of all $i \in I$ such that Δ is transverse (= not intransverse) to u_i.

(8.14) Remark. We see that for any $\Delta \subset \{\mathcal{M}^\nu\}$ there is a minimal subspace $W \subset V = \mathcal{M}/\mathcal{M}^2$ such that $\operatorname{Init}(\Delta) \subset \mathbf{k}[W]$. Moreover, Δ is intransverse to a fixed variable u_i, $i \in I$, if and only if W is intransverse to $L_i = [(u_i) \bmod \mathcal{M}^2] \subset V$.

9. Contact filtrations and their structure

(9.1) Notation. Throughout this section $\mathcal{O}$ will be a complete normal crossing ring of characteristic zero with a maximal ideal $\mathcal{M} \subset \mathcal{O}$, $\mathcal{O}/\mathcal{M} = \mathbf{k}$, and with fixed variables u_i, $i \in I$. Let $\operatorname{Der} = \operatorname{Der}_{\mathcal{O}}$ denote the module of all derivations of $\mathcal{O}$ (over $\mathbf{Q}$). For any subset $J \subset I$ let

(9.1.1) $$\operatorname{Der}_{\mathcal{O},J} = \operatorname{Der}_J = \left\{D \in \operatorname{Der} \;\middle|\; Du_i \in (u_i) \text{ for all } i \in J\right\}$$

— the module of all derivations tangent to the submanifolds $u_i = 0$, $i \in J$.

If $\{u_1, u_2, \ldots, u_\ell, v_1, v_2, \ldots, v_s\}$ is a coordinate system such that the fixed variables of $\mathcal{O}$ are $u_1, u_2, \ldots, u_\ell$, then any choice of a coefficient field $\mathbf{k} \hookrightarrow \mathcal{O}$ gives an isomorphism

$$\mathcal{O} \cong \mathbf{k}[[u_1, u_2, \ldots, u_\ell, v_1, v_2, \ldots, v_s]]$$

which shows that

(9.1.2) $$\operatorname{Der} = \sum \mathcal{O} \cdot \frac{\partial}{\partial u_i} + \sum \mathcal{O} \cdot \frac{\partial}{\partial v_i} + \operatorname{Der}(\mathbf{k}, \mathcal{O})$$

where $\operatorname{Der}(\mathbf{k}, \mathcal{O})$ is the module of all derivations from $\mathbf{k}$ to $\mathcal{O}$, and

(9.1.3) $$\operatorname{Der}_J = \sum_{i \in J} \mathcal{O} \cdot u_i \frac{\partial}{\partial u_i} + \sum_{i \notin J} \mathcal{O} \cdot \frac{\partial}{\partial u_i} + \sum \mathcal{O} \cdot \frac{\partial}{\partial v_i} + \operatorname{Der}(\mathbf{k}, \mathcal{O}) .$$

Now let Δ be an integrally closed filtration in $\mathcal{O}$ satisfying $\Delta \subset \{\mathcal{M}^\nu\}$. Recall that by (8.13) $I_{\mathrm{tr}}(\Delta)$ is the subset of all such $i \in I$ that Δ is transverse to u_i.

(9.2) Definition. *We shall say that a filtration* $\Delta \subset \{\mathcal{M}^\nu\}$ *in* $\mathcal{O}$ *is contact if it is integrally closed and satisfies the following conditions:*

(9.2.1) $\mathrm{Der}_{I_{\mathrm{tr}}(\Delta)} \cdot \Delta(\nu) \subset \Delta(\nu - 1)$ *for any* $\nu \in \mathbf{Q}_+$ *;*

(9.2.2) $u_i \in \Delta(1)$ *for any* $i \in I \setminus I_{\mathrm{tr}}(\Delta)$ *(i.e., any* u_i *that is intransverse to* Δ *lies in* $\Delta(1)$ *).*

We shall also say that (1) is a contact filtration.

(9.3) Remark. We can translate the conditions (9.2.1) and (9.2.2) into the language of the integrally closed subalgebras $R_t(\Delta) \subset \mathcal{O}[t^{\mathbf{Q}}]$. Indeed, they are equivalent to the following two conditions:

(9.3.1) $R_t(\Delta)$ is preserved under the action of $(1/t)\,\mathrm{Der}_{I_{\mathrm{tr}}(\Delta)}$, i.e.,

$$\frac{1}{t}\,\mathrm{Der}_{I_{\mathrm{tr}}(\Delta)} \cdot R_t(\Delta) \subset R_t(\Delta)$$

(here $\mathrm{Der}_{I_{\mathrm{tr}}(\Delta)}$ acts on $\mathcal{O}[t^{\mathbf{Q}}]$ coefficientwise);

(9.3.2) $tu_i \in R_t(\Delta)$ for all u_i that are intransverse to Δ (i.e., $i \in I \setminus I_{\mathrm{tr}}(\Delta)$).

(9.4) Examples. Consider $\Delta = \{\mathcal{M}^\nu\}$; then $I_{\mathrm{tr}}(\Delta) = \emptyset$, i.e., all the fixed variables are intransverse o Δ . Thus, $\mathrm{Der}_{I_{\mathrm{tr}}(\Delta)} = \mathrm{Der}$, and Δ is clearly contact, since $\mathrm{Der} \cdot \Delta(\nu) \subset \Delta(\nu - 1)$ and all $u_i \in \Delta(1) = \mathcal{M}$.

Now consider another filtration Δ' defined by $\Delta'(\nu) = \mathcal{M}^n$ where n is the minimal integer such hat $n > \nu$. In other words, $\Delta'(\nu)$ consists of all elements of $\Delta(\nu)$, $\Delta = \{\mathcal{M}^\nu\}$, whose *initial forms are ero.* We shall denote $\Delta' = \{\mathcal{M}^{\nu+0}\}$. It is easy to see that Δ' is integrally closed (although not finitely generated) and contact; $I_{\mathrm{tr}}(\Delta') = I$, i.e., all the fixed variables are transverse to Δ' .

(9.5) Proposition. *Suppose* $\mathcal{O} = \mathcal{O}_1[[x_1, x_2, \ldots, x_n]]$ *as normal crossing rings (in particular, this means that* $\mathcal{O}_1$ *is complete, since* $\mathcal{O}$ *is). If* Δ *is a contact filtration in* $\mathcal{O}_1$ *, then its suspension*

$$\{(x_1, x_2, \ldots, x_n)^\nu\} * \Delta$$

is a contact filtration in $\mathcal{O}$.

The proof is straightforward and it is left to the reader. ■

(9.6) Corollary. *The suspension of* $\{\mathcal{M}^{\nu+0}\}$ *is contact.* ■

(9.7) Remark. Let $W \subset \mathcal{M}/\mathcal{M}^2$ be a linear subspace which is weakly transverse to the fixed variables. Consider a filtration Δ such that $\Delta(\nu)$ consists of those elements of $\mathcal{M}^\nu$ such that their initial forms (of course, nonzero only for integer ν) lie in $\mathbf{k}[W]$. It is easy to see that Δ is a suspension of the filtration $\{\mathcal{M}_1^{\nu+0}\}$ in an appropriate subring $\mathcal{O}_1 \subset \mathcal{O}$. Thus, Δ is contact.

(9.8) Proposition. *For any integrally closed filtration* Δ *in* $\mathcal{O}$ *there exists a unique minimal contact filtration* Δ_0 *containing* Δ *, i.e., such that it is contact and it is contained in all other contact filtrations containing* Δ .

(9.9) Proof: First of all, note that the uniqueness of such filtration Δ_0 is obvious, if its existence is proved; thus, we have to prove the existence.

In case $\Delta \not\subset \{\mathcal{M}^\nu\}$ we can take $\Delta_0 = (1)$ — it clearly has the required properties.

Suppose $\Delta \subset \{\mathcal{M}^\nu\}$. By (9.3) we need to show the existence of a minimal subalgebra of $\mathcal{O}[t^{\mathbf{Q}}]$ which is graded, is integrally closed, contains $R_t(\Delta)$ and satisfies (9.3.1) and (9.3.2).

Indeed, let $I_1 = I_{\mathrm{tr}}(\Delta)$. Take the minimal subalgebra of $\mathcal{O}[t^{\mathbf{Q}}]$, which is graded, is integrally closed, is closed under the action of $(1/t)\,\mathrm{Der}_{I_1}$ and contains tu_i for all $i \in I \setminus I_1$; then this subalgebra has the form $R_t(\Delta_1)$ for some integrally closed filtration Δ_1. Let $I_2 = I_{\mathrm{tr}}(\Delta_1)$; then $I_2 \subset I_1$; if $I_2 \neq I_1$, then we repeat the same procedure for $R_t(\Delta_1)$ and the action of $(1/t)\,\mathrm{Der}_{I_2}$, and get a subalgebra $R_t(\Delta_2)$. If $I_3 = I_{\mathrm{tr}}(\Delta_2) \subset I_2$ and $I_3 \neq I_2$, then we repeat the same procedure again and continue this way until at some step we get a subring $R_t(\Delta_k)$ which is graded, integrally closed, and closed under the action of $(1/t)\,\mathrm{Der}_{I_k}$ where $I_k = I_{k+1} = I_{\mathrm{tr}}(\Delta_k)$. Then the filtration $\Delta_0 = \Delta_k$ is clearly the minimal contact filtration containing Δ.* ■

(9.10) Proposition. *Let Δ be a contact filtration such that $[\Delta(1) \bmod \mathcal{M}^2] = \mathrm{Im}(\Delta(1) \to \mathcal{M}/\mathcal{M}^2)$ is nonzero. Let $\mathcal{O}_1$ be a normal crossing subring of $\mathcal{O}$ and $x_1, x_2, \ldots, x_n \in \Delta(1)$ be such elements that:*

(9.10.1) the image of $x_1, x_2, \ldots, x_n$ in $\mathcal{M}/\mathcal{M}^2$ form a basis of $\Delta(1) \bmod \mathcal{M}^2$, and

(9.10.2) $\mathcal{O} = \mathcal{O}_1[[x_1, x_2, \ldots, x_n]]$ as normal crossing rings (in particular, $\mathcal{O}_1$ is complete).

Then there exists a unique contact filtration Δ' in $\mathcal{O}_1$ such that Δ is the suspension of Δ':

(9.10.3)
$$\Delta = \{(x_1, x_2, \ldots, x_n)^\nu\} * \Delta'$$

In particular, $[\Delta' \bmod \mathcal{M}^2] = 0$.

(9.11) Proof: Let

(9.11.1)
$$\Delta'(\nu) = \Delta(\nu) \cap \mathcal{O}_1 .$$

Then clearly Δ' is an integrally closed filtration in $\mathcal{O}_1$. Since $x_1, x_2, \ldots, x_n \in \Delta(1)$ and $\Delta' \subset \Delta$,

(9.11.2)
$$\Delta \supset \{(x_1, x_2, \ldots, x_n)^\nu\} * \Delta' .$$

It is also clear that if an integrally closed filtration Δ' in $\mathcal{O}_1$ satisfies (9.10.3), then Δ' is given by (9.11.1), which means that such Δ' is unique. Thus, what requires proof is that Δ' is contact and there is an equality in (9.11.2).

Note that the fixed variables of $\mathcal{O}_1$ are exactly u_i, $i \in I_1(\Delta)$. It follows from (9.11.1) that

(9.11.3)
$$[\Delta'(1) \bmod \mathcal{M}^2] = 0 .$$

Thus, all the fixed variables of $\mathcal{O}_1$ are transverse to Δ', so there are no intransverse fixed variables. This means that the condition (9.2.2) for Δ' is vacuous (and thus satisfied trivially). This also means that $I_{\mathrm{tr}}(\Delta') = I_{\mathrm{tr}}(\Delta)$, so $\mathrm{Der}_{\mathcal{O}', I_{\mathrm{tr}}(\Delta')} \subset \mathrm{Der}_{\mathcal{O}, I_{\mathrm{tr}}(\Delta)} = \mathrm{Der}_{I_{\mathrm{tr}}(\Delta)}$, which shows that (9.2.1) for Δ' is also true. Thus, Δ' is contact.

Let $f \in \Delta(\nu_1) \subset \mathcal{O} = \mathcal{O}_1[[x_1, x_2, \ldots, x_n]]$. Then

(9.11.4)
$$f = \sum_{\alpha = (\alpha_1, \alpha_2, \ldots, \alpha_n) \in \mathbf{Z}_+^n} f_\alpha x_1^{\alpha_1} x_2^{\alpha_2} \ldots x_n^{\alpha_n}$$

* We shall see later in (9.13 ii) that, in fact, the first step of this procedure is always its final step, because $I_1 = I_{\mathrm{tr}}(\Delta) = I_{\mathrm{tr}}(\Delta_0) = I_k$.

where $f_\alpha \in \mathcal{O}_1$, and we need to show that

$$f \in [\{(x_1, x_2, \ldots, x_n)^\nu\} * \Delta'](\nu_1)$$

or, in other words, that $f_\alpha \in \Delta'(\nu_1 - |\alpha|)$. This is trivial for $|\alpha| \geq \nu_1$, so we may drop all the terms with $|\alpha| \geq \nu_1$ from (9.11.4). By (9.11.2) this will preserve the property $f \in \Delta(\nu_1)$, so we may assume that the summation in (9.11.4) is taken over a finite set of $\alpha \in \mathbf{Z}_+^n$ satisfying $|\alpha| < \nu_1$.

We may also apply the induction on $\nu_1 - |\alpha| = k$ (which, however, is not necessarily an integer, if ν_1 is not) and assume that if $|\alpha| > \nu_1 - k$, then $f_\alpha \in \Delta'(\nu_1 - |\alpha|)$. Again dropping all these terms from (9.11.4), we may assume that the summation in (9.11.4) is taken only over those $\alpha \in \mathbf{Z}_+^n$ with $|\alpha| \leq \nu_1 - k$. To carry out the induction step, we have to show that $f_\alpha \in \Delta'(\nu - |\alpha|)$ for such α that $|\alpha| = \nu_1 - k$.

Indeed, $\partial/\partial x_i \in \mathrm{Der}_{I_1(\Delta)}$ for all $i = 1, 2, \ldots, n$, and since f now is a polynomial in $x_1, x_2, \ldots, x_n$ of degree at most $\nu_1 - k$,

$$f_\alpha = \frac{1}{\alpha_1!\alpha_2!\ldots\alpha_n!} \frac{\partial^{\nu_1 - k}}{\partial x_1^{\alpha_1} \partial x_2^{\alpha_2} \ldots \partial x_n^{\alpha_n}} f$$

for each α satisfying $|\alpha| = \nu_1 - k$. Each application of $\partial/\partial x_1$ moves us from $\Delta(\nu)$ to $\Delta(\nu - 1)$, so we see that $f_\alpha \in \Delta(k) = \Delta(\nu_1 - |\alpha|)$. Finally, $f_\alpha \in \mathcal{O}_1$, so

$$f_\alpha \in \Delta(\nu_1 - |\alpha|) \cap \mathcal{O}_1 = \Delta'(\nu_1 - |\alpha|)$$

which completes the proof. ■

(9.12) Corollary. *If Δ is a contact filtration in $\mathcal{O}$, then $\mathrm{Init}(\Delta)$ is a polynomial subalgebra in $\mathrm{gr}\,\mathcal{O}$ generated by $\Delta(1) \bmod \mathcal{M}^2$, and $[\Delta(1) \bmod \mathcal{M}^2] \subset \mathcal{M}/\mathcal{M}^2 = \mathrm{gr}_1\,\mathcal{O}$ is weakly transverse to the fixed variables.*

The proof is a direct application of (9.10). ■

(9.13) Proposition. *Let Δ be an integrally closed filtration in $\mathcal{O}$, and let Δ_0 be the minimal contact filtration containing it.*

(i) If $\Delta \not\subset \{\mathcal{M}^\nu\}$, then $\Delta_0 = (1)$.

(ii) If $\Delta \subset \{\mathcal{M}^\nu\}$, then $\Delta_0 \subset \{\mathcal{M}^\nu\}$ and $\mathrm{Init}(\Delta_0)$ is the minimal polynomial subalgebra of $\mathrm{gr}\,\mathcal{O}$ having the following properties:

(9.13.1) it is generated by a subspace $W \subset \mathrm{gr}_1\,\mathcal{O} = \mathcal{M}/\mathcal{M}^2$;

(9.13.2) this subspace W is weakly transverse to the fixed variables;

(9.13.3) the subalgebra $\mathrm{Init}(\Delta_0)$ contains $\mathrm{Init}(\Delta)$.

(9.14) Proof: Part (i) follows from the definitions.

Let us look at (ii). Suppose $\Delta \subset \{\mathcal{M}^\nu\}$. Clearly, $\{\mathcal{M}^\nu\}$ is a contact filtration and $\Delta \subset \{\mathcal{M}^\nu\}$; thus, by minimality of Δ_0 we get $\Delta_0 \subset \{\mathcal{M}^\nu\}$.

Consider $\mathrm{Init}(\Delta) \subset \mathrm{gr}\,\mathcal{O} = \mathbf{k}[V]$ where $V = \mathcal{M}/\mathcal{M}^2$. By (8.1) and (8.6) there exists a unique minimal subspace $W \subset V$ which is weakly transverse to the fixed variables and such that $\mathrm{Init}(\Delta) \subset \mathbf{k}[W]$. Clearly, $\mathbf{k}[W]$ is the minimal polynomial subalgebra of $\mathrm{gr}\,\mathcal{O}$ having the properties (9.13.1)–(9.13.3).

At the same time (9.12) clearly shows that $\mathrm{Init}(\Delta_0)$ also has the properties (9.13.1)–(9.13.3); thus, the minimality of $\mathbf{k}[W]$ yields $\mathbf{k}[W] \subset \mathrm{Init}(\Delta_0)$, and we have to show that $\mathrm{Init}(\Delta_0) \subset \mathbf{k}[W]$.

Let Δ_1 be the filtration of Remark 9.7 whose ν-th term consists of those elements of $\mathcal{M}^\nu$ whose initial forms lie in $\mathbf{k}[W]$. Clearly, $\Delta \subset \Delta_1$. However, by (9.7) Δ_1 is contact; thus, $\Delta_0 \subset \Delta_1$ and $\mathrm{Init}(\Delta_0) \subset \mathrm{Init}(\Delta_1) = \mathbf{k}[W]$, and this completes the proof of (ii). ∎

For any positive integer ν, denote by Diff^ν the module of *differential operators of order* $\leq \nu$ acting on the ring $\mathcal{O}$; they are linear operators on $\mathcal{O}$ which are polynomials of degree $\leq \nu$ in the elements of Der. Similarly, denote by Diff^ν_J (here $J \subset I$) the module of all differential operators which are polynomials of degree $\leq \nu$ in the elements of Der_J. Clearly, $\mathrm{Diff}^\nu_J \subset \mathrm{Diff}^\nu$.

(9.15) Proposition. *Let Δ and Δ_0 be as in (9.13 ii). Assume also that Δ is generated by a family f_α, $\alpha \in A$, with the weights ν_α, $\alpha \in A$. Then the subspace $[\Delta_0(1) \bmod \mathcal{M}^2] \subset \mathcal{M}/\mathcal{M}^2$ can be characterized in the following way. Consider the ideal*

$$\mathcal{B} = \sum_{\{\alpha \in A \mid \nu_\alpha \in \mathbf{Z}_+\}} \mathrm{Diff}^{\nu_\alpha - 1}_{I_{\mathrm{tr}}(\Delta)} f_\alpha + \sum_{i \in I \setminus I_{\mathrm{tr}}(\Delta)} (u_i)\,. \tag{9.15.1}$$

Then $\mathcal{B} \neq (1)$ and $[\mathcal{B} \bmod \mathcal{M}^2] = [\Delta_0(1) \bmod \mathcal{M}^2]$.

(9.16) Proof: The assumptions of (9.13 ii) include $\Delta \subset \{\mathcal{M}^\nu\}$; thus, $f_\alpha \in \mathcal{M}^n$, where n is the minimal integer such that $n \geq \nu_\alpha$, and this shows that $\mathcal{B} \neq (1)$.

Δ_0 being contact clearly yields $\mathcal{B} \subset \Delta_0(1)$; thus,

$$[\mathcal{B} \bmod \mathcal{M}^2] \subset [\Delta_0(1) \bmod \mathcal{M}^2]$$

and we have to prove the opposite inclusion.

It is also not hard to see (cf. (8.2) and (8.9)) that the initial forms $[f_\alpha \bmod \mathcal{M}^{\nu_\alpha+1}]$ (for such α that $\nu_\alpha \in \mathbf{Z}_+$) lie in $\mathbf{k}[W]$, where $W = [\mathcal{B} \bmod \mathcal{M}^2]$.

From $W \subset [\Delta_0(1) \bmod \mathcal{M}^2]$ we see that all the variables which are intransverse to W, are intransverse to $\Delta_0(1) \bmod \mathcal{M}^2$, or equivalently, to Δ_0 and Δ. On the other hand, (9.15.1) shows that the variables which are intransverse to Δ, are intransverse to W. Thus, the sets of fixed variables which are intransverse to W, Δ_0, and Δ, coincide.

Now we see from (9.15.1) that W is weakly transverse to the fixed variables. As in (9.14), let Δ_1 be the subfiltration of $\{\mathcal{M}^\nu\}$ consisting of all elements whose initial forms lie in $\mathbf{k}[W]$ (see (9.7)). Then $f_\alpha \in \Delta_1(\nu_\alpha)$ for all α, so $\Delta \subset \Delta_1$.

By (9.7) Δ_1 is contact, so $\Delta_0 \subset \Delta_1$, and $[\Delta_0(1) \bmod \mathcal{M}^2] \subset [\Delta_1(1) \bmod \mathcal{M}^2] = W = [\mathcal{B} \bmod \mathcal{M}^2]$, and this completes the proof. ∎

(9.17) Remark. The reader familiar with the theory of maximal contact of [Giraud 1], will recognize the similarity between the term

$$\sum \mathrm{Diff}^{\nu_\alpha - 1}_{I_{\mathrm{tr}}(\Delta)} f_\alpha$$

in the formula (9.15.1), and the formulas 3.1.3 (1), (2) of [Giraud 1]. It is this similarity that accounts for the term "contact filtration". Cf. also [Youssin 1], (5.2).

10. Galois extensions

(10.1). In Section 6 we explained how to multiply and divide a filtration by a fractional monomial — we should raise this fractional monomial to a high power so that it becomes a "genuine" monomial, and then multiply (or divide) the appropriate term of the filtration by this monomial.

However, it is much more natural to consider a fractional monomial as an element of some extension of the original ring — in fact, a finite integral Galois extension. (We say that $\mathcal{O} \subset \mathcal{O}'$ is an *integral Galois extension* if $\mathcal{O}$ is an integrally closed domain and $\mathcal{O}'$ is its integral closure in a Galois extension of the quotient field.)

In this section we show how integrally closed filtrations behave under integral Galois extensions of the ambient ring.

(10.2) Remark. Given a ring homomorphism $\pi : \mathcal{O} \to \mathcal{O}'$ and the integrally closed filtration Δ in $\mathcal{O}$, we defined in (4.6) an integrally closed filtration $\pi_*\Delta$ as the minimal filtration in $\mathcal{O}'$ containing $\pi(\Delta(\nu))$. Given an integrally closed filtration Δ' in $\mathcal{O}'$, we can define the filtration $\pi^{-1}\Delta'$ by

$$(\pi^{-1}\Delta')(\nu) = \pi^{-1}(\Delta'(\nu))$$

Clearly, $\pi^{-1}\Delta'$ is integrally closed. (In case π is an inclusion, we shall sometimes write $\pi^{-1}\Delta' = \Delta' \cap \mathcal{O}$.)

One may ask whether the operations π_* and π^{-1} are inverse to each other, and of course, in general they are not.

(10.3) Proposition. *Let $\pi : \mathcal{O} \hookrightarrow \mathcal{O}'$ be an integral Galois extension of integral domains, and let Δ be an integrally closed filtration in $\mathcal{O}$. Then*

$$\pi^{-1}\pi_*\Delta = \Delta$$

(10.4) Proof: To simplify the notation, we shall identify $\mathcal{O}$ with $\pi(\mathcal{O})$. Let

$$\Delta'(\nu) = \pi(\Delta(\nu)) \cdot \mathcal{O}' = \Delta(\nu) \cdot \mathcal{O}' .$$

Then Δ' is a decreasing filtration in $\mathcal{O}'$ by ideals numbered by nonnegative rationals, and it satisfies the multiplicativity properties (2.1.1) and (2.1.2). By definition $\pi_*\Delta$ is the minimal integrally closed filtration in $\mathcal{O}'$ containing Δ'; thus, by (3.8) any $f \in (\pi_*\Delta)(\nu)$ satisfies an equation

$$f^n + g_1 f^{n-1} + g_2 f^{n-2} + \ldots + g_n = 0 \qquad (10.4.1)$$

where $g_i \in \Delta'(i\nu) = \Delta(i\nu) \cdot \mathcal{O}'$.

Take any $f \in (\pi^{-1}\pi_*\Delta)(\nu) = (\pi_*\Delta)(\nu) \cap \mathcal{O}$; we have to show $f \in \Delta(\nu)$. Since $f \in (\pi_*\Delta)(\nu)$, f should satisfy an equation (10.4.1) with $g_i \in \Delta(i\nu) \cdot \mathcal{O}'$.

Now note that we did not assume $\mathcal{O}'$ to be a finite extension of $\mathcal{O}$; however, we can always take a finite integral Galois extension $\mathcal{O}''$, $\mathcal{O} \subset \mathcal{O}'' \subset \mathcal{O}'$, such that $g_1, g_2, \ldots, g_n \in \mathcal{O}''$. Then $\mathrm{Gal}(\mathcal{O}''/\mathcal{O})$ is finite, and an application of any $\sigma \in \mathrm{Gal}(\mathcal{O}''/\mathcal{O})$ to (10.4.1) yields

$$f^n + \sigma(g_1) f^{n-1} + \sigma(g_2) f^{n-2} + \ldots + \sigma(g_n) = 0 . \qquad (10.4.2)$$

(Note that $\sigma(f)=f$ as $f\in\mathcal{O}$.) Multiplying the equations (10.4.2) for all $\sigma\in \mathrm{Gal}(\mathcal{O}''/\mathcal{O})$, we get an equation of the form

$$f^{Nn}+h_1f^{Nn-1}+h_2f^{Nn-2}+\dots h_{Nn}=0 \tag{10.4.3}$$

where $h_i\in\Delta'(i\nu)=\Delta(i\nu)\cdot\mathcal{O}'$ and N is the number of elements in $\mathrm{Gal}(\mathcal{O}''/\mathcal{O})$. However, the equation (10.4.3) is preserved under the action of Gal $(\mathcal{O}''/\mathcal{O})$; thus, $\sigma(h_i)=h_i$ for all $\sigma\in\mathrm{Gal}(\mathcal{O}''/\mathcal{O})$, so $h_i\in\mathcal{O}$.

Thus, $h_i\in\Delta(i\nu)\cdot\mathcal{O}'$ and $h_i\in\mathcal{O}$. Let

$$h_i=\sum a_sb_s \tag{10.4.4}$$

where $a_s\in\Delta(i\nu),\ b_s\in\mathcal{O}'$. Then for any σ ,

$$h_i=\sigma(h_i)=\sum a_s\cdot\sigma(b_s)\ . \tag{10.4.5}$$

Multiplying (10.4.5) for all $\sigma\in\mathrm{Gal}(\mathcal{O}''/\mathcal{O})$, we get

$$h_i^N=\sum\widetilde{a}_s\widetilde{b}_s$$

where $\widetilde{a}_s\in\Delta(Ni\nu)$, $\widetilde{b}_s\in\mathcal{O}$. Thus, $h_i^N\in\Delta(Ni\nu)$, and the integral closedness of Δ gives successively $h_i\in\Delta(i\nu)$ and $f\in\Delta(\nu)$. ■

(10.5) Remark. Let $\pi:\mathcal{O}\to\mathcal{O}'$ be a ring homomorphism, and assume $\mathcal{O}$ and $\mathcal{O}'$ are normal crossing rings. If u^α is a fractional monomial in $\mathcal{O}$, it may happen that $\pi(u^\alpha)$ may be considered as a fractional monomial in $\mathcal{O}'$; a condition for this is that the image $\pi(u_i)$ of any fixed variable $u_i\in\mathcal{O}$ is either invertible or a *power* (of course, an integer power) of a fixed variable in $\mathcal{O}'$.

If this is the case and if Δ is an integrally closed filtration in $\mathcal{O}$, then we can compare the two filtrations $\pi_*(u^\alpha\cdot\Delta)$ and $\pi(u^\alpha)\cdot\pi_*\Delta$.

(10.6) Proposition. *If $\pi:\mathcal{O}\to\mathcal{O}'$ is as in (10.5) and Δ is an integrally closed filtration in $\mathcal{O}$, then*

$$\pi_*(u^\alpha\cdot\Delta)=\pi(u^\alpha)\cdot\pi_*\Delta\ .$$

(10.7) Proof: $\pi_*(u^\alpha\cdot\Delta)$ is the minimal integrally closed filtration whose ν-th term contains $\pi\big(u^\beta\cdot\Delta(\nu)\big)$, where u^β is the minimal integer ("genuine") monomial divisible by u^α . Since

$$\pi\big(u^\beta\cdot\Delta(\nu)\big)\subset[\pi(u^\alpha)\cdot\pi_*\Delta](\nu)$$

we see that

$$\pi_*(u^\alpha\cdot\Delta)\subset[\pi(u^\alpha)\cdot\pi_*\Delta]\ .$$

To see the opposite inclusion, take ν such that $u^{\nu\alpha}$ is an integer monomial, and take $f\in[\pi(u^\alpha)\cdot\pi_*\Delta](\nu)$. By (6.9.1), $f=\pi(u^{\nu\alpha})\cdot\widetilde{f}$, where $\widetilde{f}\in(\pi_*\Delta)(\nu)$. Then, as in (10.4), $\widetilde{f}$ satisfies an equation

$$\widetilde{f}^n+\widetilde{g}_1\widetilde{f}^{n-1}+\widetilde{g}_2\widetilde{f}^{n-1}+\dots+\widetilde{g}_n=0 \tag{10.7.2}$$

with $\widetilde{g}_i\in\pi\big(\Delta(i\nu)\big)\cdot\mathcal{O}'$. Consequently, f satisfies the equation

$$f^n+g_1f^{n-1}+g_2f^{n-2}+\dots+g_n=0 \tag{10.7.3}$$

where $g_i = \pi(u^{i\nu\alpha}) \cdot \widetilde{g}_i \in \pi(u^{i\nu\alpha}) \cdot \pi(\Delta(i\nu)) \cdot \mathcal{O}' = \pi(u^{i\nu\alpha} \cdot \Delta(i\nu)) \cdot \mathcal{O}' = \pi[(u^\alpha \cdot \Delta)(i\nu)] \cdot \mathcal{O}'$. Thus, $f \in [\pi_*(u^\alpha \cdot \Delta)](\nu)$, and by integral closedness $\pi(u^\alpha) \cdot \pi_*\Delta \subset \pi_*(u^\alpha \cdot \Delta)$. ∎

(10.8) Remark. Propositions 10.3 and 10.6 show that we may indeed understand the product of a filtration by a fractional monomial via the ring extensions: for any $\nu \in \mathbf{Q}_+$ take the ring extension $\pi : \mathcal{O} \hookrightarrow \mathcal{O}'$ which makes $\pi(u^{\nu\alpha})$ an integer ("genuine") monomial; then

(10.8.1) $$(u^\alpha \cdot \Delta)(\nu) = \left[u^{\nu\alpha} \cdot [(\pi_*\Delta)(\nu)]\right] \cap \mathcal{O}$$

i.e., to find out what $(u^\alpha \cdot \Delta)(\nu)$ is, we can take the pushforward $\pi_*\Delta$ into an extension where $\pi(u^{\nu\alpha})$ becomes a "genuine" monomial, take its ν-th term $(\pi_*\Delta)(\nu)$, multiply it (as an ideal, not as a filtration!) by $\pi(u^{\nu\alpha})$, and finally take the intersection with $\mathcal{O}$.

Note that in (10.6) we did not assume much about the homomorphism $\pi : \mathcal{O} \to \mathcal{O}'$; in paricular, we did not assume it to be an integral ring extension.

Of course, we cannot expect any similar result to be true for the operation of division of a filtration by a monomial, since even division of ideals does not behave well under the homomorphisms of the ambient ring. Namely, for a general ring homomorphism $\pi : \mathcal{O} \to \mathcal{O}'$ and an ideal $\mathcal{M} \subset \mathcal{O}$ the ideal $\pi(\mathcal{M} : u)$ does not have to coincide with $\pi(\mathcal{M}) : \pi(u)$. However, in the case of integral Galois extensions we have the following

(10.9) Proposition. *Let $\pi : \mathcal{O} \hookrightarrow \mathcal{O}'$ be an integral Galois extension of normal crossing rings. Assume that for each fixed variable u_i in $\mathcal{O}$, $\pi(u_i)$ is an integer power of a fixed variable in $\mathcal{O}'$ or invertible. Let Δ be an integrally closed filtration in $\mathcal{O}$ and u^α a fractional monomial in $\mathcal{O}$. Then*

$$\pi_*(\Delta : u^\alpha) = (\pi_*\Delta) : \pi(u^\alpha) .$$

(10.10) Proof: As in (10.7), we notice immediately that

(10.10.1) $$\pi_*(\Delta : u^\alpha) \subset \pi_*\Delta : \pi(u^\alpha)$$

for any ring homomorphism π, and it is for the opposite inclusion that we have to use the assumption that π is an integral Galois extension.

Again, like in (10.7), take ν such that $u^{\nu\alpha}$ is an integer monomial and $f \in \left[\pi_*\Delta : \pi(u^\alpha)\right](\nu)$. Let $\widetilde{f} = \pi(u^{\nu\alpha}) \cdot f$; by (6.9.2) $\widetilde{f} \in (\pi_*\Delta)(\nu)$.

Now identify $\mathcal{O}$ with $\pi(\mathcal{O})$ and take a finite integral Galois extension $\mathcal{O}''$, such that $\mathcal{O} \subset \mathcal{O}'' \subset \mathcal{O}'$ and $f, \widetilde{f} \in \mathcal{O}''$.

We know that $\widetilde{f} \in (u^{\nu\alpha} \cdot \mathcal{O}') \cap \mathcal{O}''$ and $\widetilde{f} \in (\pi_*\Delta)(\nu) \cap \mathcal{O}''$. It is easy to see that both $(u^{\nu\alpha} \cdot \mathcal{O}') \cap \mathcal{O}''$ and $(\pi_*\Delta)(\nu) \cap \mathcal{O}''$ are preserved under the action of $\mathrm{Gal}(\mathcal{O}''/\mathcal{O})$; thus, for any $\sigma \in \mathrm{Gal}(\mathcal{O}''/\mathcal{O})$, $\sigma(\widetilde{f}) \in u^{\nu\alpha} \cdot \mathcal{O}'$ and $\sigma(\widetilde{f}) \in (\pi_*\Delta)(\nu)$. Let n be the number of elements in $\mathrm{Gal}(\mathcal{O}''/\mathcal{O})$, and consider the elementary symmetric polynomials $\widetilde{g}_1, \widetilde{g}_2, \ldots, \widetilde{g}_n$ in all the conjugates $\sigma(\widetilde{f})$, $\sigma \in \mathrm{Gal}(\mathcal{O}''/\mathcal{O})$. Clearly, $\widetilde{g}_i \in \mathcal{O}$, $\widetilde{g}_i \in u^{i\nu\alpha} \cdot \mathcal{O}'$, $\widetilde{g}_i \in (\pi_*\Delta)(i\nu)$, and $\widetilde{f}$ satisfies the equation

(10.10.2) $$\widetilde{f}^n + \widetilde{g}_1 \widetilde{f}^{n-1} + \widetilde{g}_2 \widetilde{f}^{n-2} + \ldots + \widetilde{g}_n = 0 .$$

Note that $\widetilde{g}_i \in (\pi_*\Delta)(i\nu) \cap \mathcal{O} = (\pi^{-1}\pi_*\Delta)(i\nu) = \Delta(i\nu)$ by (10.3).

Let $g_i = \widetilde{g}_i / u^{i\nu\alpha}$; as $\widetilde{g}_i \in u^{i\nu\alpha} \cdot \mathcal{O}'$, $g_i \in \mathcal{O}'$, so g_i is the integral over $\mathcal{O}$. At the same time g_i clearly lies in the quotient field of $\mathcal{O}$. As $\mathcal{O}$ is a normal crossing ring, it is a regular local ring, so it is integrally closed in its quotient field. Thus, $g_i \in \mathcal{O}$, and consequently $g_i \in \Delta(i\nu) : u^{i\nu\alpha}$.

By (6.9.2), $\Delta(i\nu):u^{i\nu\alpha}=(\Delta:u^{\alpha})(i\nu)$, so $g_i\in(\Delta:u^{\alpha})(i\nu)$. Since $\widetilde{f}=u^{\nu\alpha}\cdot f$, f satisfies the equation

(10.10.3)
$$f^n+g_1f^{n-1}+g_2f^{n-2}+\ldots+g_n=0\ .$$

Here $f\in\mathcal{O}'$ and $g_i\in(\Delta:u^{\alpha})(i\nu)\subset[\pi_*(\Delta:u^{\alpha})](i\nu)$, so $f\in[\pi_*(\Delta:u^{\alpha})](\nu)$.

Thus, for any ν such that $\nu\alpha$ is integer,

(10.10.4)
$$[\pi_*(\Delta:u^{\alpha})](\nu)\supset[\pi_*\Delta:\pi(u^{\alpha})](\nu)$$

so, by integral closedness, $\pi_*(\Delta:u^{\alpha})\supset\pi_*\Delta:\pi(u^{\alpha})$. ∎

(10.11) Remark. Proposition 10.9 shows that we may understand the quotients of filtrations by fractional monomials in the same way as the products where understood in (10.8). Namely, to find out what $(\Delta:u^{\alpha})(\nu)$ is, we have to take an integral Galois extension $\pi:\mathcal{O}\hookrightarrow\mathcal{O}'$ such that $\pi(u^{\nu\alpha})$ is a "genuine" monomial, take the pushforward $\pi_*\Delta$ of Δ into $\mathcal{O}'$, divide its ν-th term by $\pi(u^{\nu\alpha})$ (as an ideal, not as a filtration), and finally intersect the result with $\mathcal{O}$, i.e.,

(10.11.1)
$$(\Delta:u^{\alpha})(\nu)=[(\pi_*\Delta)(\nu):\pi(u^{\nu\alpha})]\cap\mathcal{O}\ .$$

11. Stably contact filtrations

Here we introduce another property of integrally closed filtrations — being stably contact — which is more restrictive than being contact, and in Section 12 we shall show that special filtrations are stably contact (at least up to a rescaling).

As in Section 9, let $\mathcal{O}$ be a complete normal crossing ring of characteristic zero with the maximal ideal $\mathcal{M}\subset\mathcal{O}$, $\mathcal{O}/\mathcal{M}=\mathbf{k}$, and with the fixed variables u_i , $i\in I$.

(11.1) Remark. Consider the extension $\mathcal{O}[\sqrt{u}]$ of $\mathcal{O}$ by all the roots (of all degrees) of all the fixed variables.

One should note here that each of the fixed variables is defined up to the multiplication by an invertible element of $\mathcal{O}$; however, the assumptions on $\mathcal{O}$ guarantee that the extension of $\mathcal{O}$ by the roots of u_i coincides with the extension by the roots of fu_i , if $f\in\mathcal{O}$ is invertible. Thus, $\mathcal{O}[\sqrt{u}]$ depends only on the normal crossing divisor in $\operatorname{Spec}\mathcal{O}$ and not on the choice of the equations u_i of its components.

Clearly, $\mathcal{O}[\sqrt{u}]$ is an integral Galois extension of $\mathcal{O}$; it is the maximal extension ramified only at that normal crossing divisor of $\operatorname{Spec}\mathcal{O}$ which is given by the structure of the normal crossing ring on $\mathcal{O}$.

Note that $\mathcal{O}[\sqrt{u}]$ is neither complete nor a Noetherian local ring (and thus not a normal crossing ring). However, $\mathcal{O}[\sqrt{u}]$ is a direct limit of complete regular local rings with compatible (as in (10.5)) structures of normal crossing rings on them.

We may apply the theory of Section 10 to the extension $\pi:\mathcal{O}\hookrightarrow\mathcal{O}[\sqrt{u}]=\widehat{\mathcal{O}}$. Given an integrally closed filtration Δ in $\mathcal{O}$, we may construct the integrally closed filtration $\pi_*\Delta$ in $\widehat{\mathcal{O}}=\mathcal{O}[\sqrt{u}]$, and by (10.3), $\pi^{-1}\pi_*\Delta=\Delta$. Thus, there is a correspondence between integrally closed filtrations in $\mathcal{O}$ and (some) integrally closed filtrations in $\widehat{\mathcal{O}}$.

Note that the derivations of $\mathcal{O}$ extend to $\widehat{\mathcal{O}}$ as rational derivations, i.e., the derivations of the quotient field of $\widehat{\mathcal{O}}$; as derivations of $\widehat{\mathcal{O}}$, they are not everywhere defined. (Note, however, that the elements of $\mathrm{Der}_{\mathcal{O},I}$ act as regular — everywhere defined — derivations on $\widehat{\mathcal{O}}$.)

Now suppose a filtration Δ in $\mathcal{O}$ is contact. We may say that this means that the derivations of $\mathcal{O}$ act on Δ in a certain way, and we may then ask whether the same derivations of $\mathcal{O}$ act in some good way on $\pi_*\Delta$. Unfortunately, the construction of $\pi_*\Delta$ involves solving algebraic equations, and this operation interacts with the derivations in a very bad way.

(11.2) Proposition. *For any integrally closed filtration Δ there exists a minimal integrally closed filtration $\widehat{\Delta}$ in $\widehat{\mathcal{O}}=\mathcal{O}[\sqrt{u}]$ such that $\widehat{\Delta}\supset\pi_*\Delta$ and $\widehat{\Delta}$ satisfies the following property:*

(11.2.1) If $f\in\widehat{\Delta}(\nu)$, $D\in\mathrm{Der}_{\mathcal{O},I_{\mathrm{tr}}(\Delta)}$ are such that $Df\in\widehat{\mathcal{O}}$ (here D is continued to $\widehat{\mathcal{O}}$ as a rational derivation), then $Df\in\widehat{\Delta}(\nu-1)$.

The proof is quite similar to (9.9) and it is left to the reader. ■

(11.3) Definition. *We shall say a filtration Δ in $\mathcal{O}$ is stably contact if it is contact and the following condition is satisfied. Let $\widehat{\Delta}$ be as in (11.2); then $\pi^{-1}\widehat{\Delta}=\Delta$.*

(11.4) Examples. One can easily see that both filtrations $\{\mathcal{M}^{\nu}\}$ and $\{\mathcal{M}^{\nu+0}\}$ (see (9.4)) are stably contact. Indeed, if $\Delta=\{\mathcal{M}^{\nu}\}$ or $\Delta=\{\mathcal{M}^{\nu+0}\}$, then in the notation of (11.1)–(11.2) $\widehat{\Delta}=\pi_*\Delta$ and it may be described in the following way. Identify $\mathcal{O}$ with a formal power series ring; then $\widehat{\mathcal{O}}=\mathcal{O}[\sqrt{u}]$ is identified with the formal power series ring where fractional powers of u_i , $i\in I$, are allowed, on the only condition that the exponents of all the monomials in each power series have a common denominator. Then $\widehat{\Delta}(\nu)$ consists of those power series that involve only the (fractional) monomials* of total degree $\geq\nu$ (for $\{\mathcal{M}^{\nu}\}$), or $>\nu$ (for $\{\mathcal{M}^{\nu+0}\}$). Clearly, in both cases $\pi^{-1}\widehat{\Delta}=\Delta$.

One should also note that (1) and (0) are also stably contact.

Another important example is the filtration $\Delta=u^{\alpha}\cdot(1)$ with $|\alpha|>1$; it is clearly a contact filtration. In this case $I_{\mathrm{tr}}(\Delta)=I$, i.e., Δ is transverse to all the fixed variables, and $\pi_*\Delta=u^{\alpha}\cdot(1)$, i.e., $(\pi_*\Delta)(\nu)=u^{\nu\alpha}\cdot\widehat{\mathcal{O}}$. Thus, $\mathrm{Diff}_{I_{\mathrm{tr}}(\Delta)}\cdot(\pi_*\Delta)(\nu)\subset(\pi_*\Delta)(\nu)\subset(\pi_*\Delta)(\nu-1)$, which shows that $\widehat{\Delta}=\pi_*\Delta$. Since $\pi^{-1}\widehat{\Delta}=\Delta$ and Δ is contact, Δ is stably contact.

(11.5) Proposition. *Let $\mathcal{O}=\mathcal{O}_1[[x_1,x_2,\dots,x_n]]$ be an equality of normal crossing rings. If Δ_1 is a stably contact filtration in $\mathcal{O}_1$, then its suspension*

$$\{(x_1,x_2,\dots,x_n)^{\nu}\}*\Delta_1$$

is a stably contact filtration in $\mathcal{O}$.

For the proof, see Appendix A5.

(11.6) Corollary. *Let $W\subset\mathcal{M}/\mathcal{M}^2$ be a linear subspace which is weakly transverse to the fixed variables. Consider the filtration Δ consisting of those elements of $\{\mathcal{M}^{\nu}\}$ whose initial forms lie in $\mathbf{k}[W]$ (see (9.7)). Then Δ is stably contact.*

---

* Here monomials are just the terms in the power series expansion of each element of $\widehat{\mathcal{O}}=\mathcal{O}[\sqrt{u}]$; thus, they may involve fractional powers of u_i, $i\in I$, and integer powers of all the other variables.

(Indeed, we have seen in (9.7) that Δ is a suspension of $\{\mathcal{M}^{\nu+0}\}$.) ■

(11.7) Proposition. *Given a contact filtration Δ in $\mathcal{O}$, there exists a unique minimal stably contact filtration Δ_0 which is contained in any other stably contact filtration containing Δ .*

(11.8) Proof: Consider the filtration $\widehat{\Delta}$ of (11.1). Clearly, $\Delta_0 = \pi^{-1}\widehat{\Delta}$ is stably contact, and it is minimal among all stably contact filtrations containing Δ . ■

(11.9) Corollary. *Given any integrally closed filtration Δ in $\mathcal{O}$, there exists a unique minimal stably contact filtration Δ_0 which is contained in any other stably contact filtration containing Δ .*

(11.10) Proof: By (9.8) there is a minimal filtration Δ_0' among the contact filtrations containing Δ , and by (11.7) there is a minimal filtration Δ_0 among the stably contact filtrations containing Δ_0' . Clearly, Δ_0 is the minimal among the stably contact filtrations containing Δ . ■

(11.11) Proposition. *Let Δ be an integrally closed filtration in $\mathcal{O}$, Δ_0' and Δ_0 respectively the minimal contact and the minimal stably contact filtrations containing Δ . Then $[\Delta_0 \bmod \mathcal{M}^2] = [\Delta_0' \bmod \mathcal{M}^2]$; thus, $\mathrm{Init}(\Delta_0)$ is described by (9.13) and (9.15).*

(11.12) Proof: All we have to show is that if Δ_0' is contact and Δ_0 is the minimal stably contact filtration containing it, then $[\Delta_0 \bmod \mathcal{M}^2] = [\Delta_0' \bmod \mathcal{M}^2]$.

We know from (9.12) that $W = [\Delta_0' \bmod \mathcal{M}^2]$ is weakly transverse to the fixed variables. Let Δ_1 be the subfiltration of $\{\mathcal{M}^\nu\}$ consisting of those elements whose initial forms lie in $\mathbf{k}[W]$ (see (9.7)). By (11.6) Δ_1 is stably contact. Clearly, $\Delta_0' \subset \Delta_1$; thus, $\Delta_0 \subset \Delta_1$ and $[\Delta_0 \bmod \mathcal{M}^2] \subset [\Delta_1 \bmod \mathcal{M}^2] = W$.

However, $\Delta_0' \subset \Delta_0$; thus, $W = [\Delta_0' \bmod \mathcal{M}^2] \subset [\Delta_0 \bmod \mathcal{M}^2]$. So we see that $[\Delta_0 \bmod \mathcal{M}^2] = [\Delta_0' \bmod \mathcal{M}^2]$. ■

(11.13) Proposition. *In the notation and the assumptions of (9.10), assume also that Δ is stably contact. Then the filtration Δ' (the one whose suspension is Δ) is also stably contact.*

(11.14) Proof: By (9.11.1), $\Delta'(\nu) = \Delta(\nu) \cap \mathcal{O}_1$.

Let $\widehat{\mathcal{O}}_1 = \mathcal{O}_1[\sqrt{u}]$ be the extension of $\mathcal{O}_1$ by all the roots (of all degrees) of all the fixed variables of $\mathcal{O}_1$. Then $\widehat{\mathcal{O}}_1 \subset \widehat{\mathcal{O}}$ and (cf. (A5.4)) $\widehat{\mathcal{O}}$ is a formal power series ring over $\widehat{\mathcal{O}}_1$ in the variables $x_1, x_2, \ldots, x_n$, where we allow fractional powers of those of $x_1, x_2, \ldots, x_n$ which are fixed variables, subject to the restrictions (A5.4.1) and (A5.4.2).

Let $\widehat{\Delta}, \widehat{\Delta}'$ be the filtrations in $\widehat{\mathcal{O}}, \widehat{\mathcal{O}}_1$ respectively, which are minimal, integrally closed, satisfy (11.2.1), and contain Δ, Δ' . Then Lemma A5.7 says that

$$\widehat{\Delta}(\nu) = \sum_{\alpha_1,\alpha_2,\ldots,\alpha_n} x_1^{\alpha_1} x_2^{\alpha_2} \ldots x_n^{\alpha_n} \widehat{\Delta}'(\nu - \textstyle\sum \alpha_i) \tag{11.14.1}$$

where fractional α_i are allowed provided $x_i^{\alpha_i}$ makes sense in $\widehat{\mathcal{O}}$, and infinite sums are allowed within the bounds of (A5.4.1) and (A5.4.2).

Since Δ is stably contact, $\widehat{\Delta} \cap \mathcal{O} = \Delta$. Comparing this with (11.14.1), we see that $\widehat{\Delta}' \cap \mathcal{O}_1 = \Delta'$,which means that Δ' is stably contact. ■

(11.15) Remark. We defined stably contact filtrations as contact filtrations satisfying the additional property $\pi^{-1}\widehat{\Delta} = \Delta$ as in (11.3). However, this additional property yields one of the properties that define a contact filtration, namely (9.2.1). Thus, we could have defined the stably contact filtrations as the ones that satisfy(9.2.2) and the additional property as in (11.3).

12. Special filtrations are stably contact

As in Sections 9 and 11, let $\mathcal{O}$ be a complete normal crossing ring of characteristic zero with the maximal ideal $\mathcal{M}$ and the fixed variables u_i, $i \in I$.

Here we prove:

(12.1) Proposition. *Special filtrations Δ satisfying $\Delta \subset \{\mathcal{M}^\nu\}$ are stably contact. In particular, they are contact.*

(12.2) Remark. For a special filtration Δ given by (7.2.3), $\Delta \subset \{\mathcal{M}^\nu\}$ is clearly equivalent to $c_1 \geq 1$. The really interesting case is $c_1 = 1$ for the following reasons. Any special filtration given by (7.2.3) with $c_1 > 1$ can be obtained by rescaling another special filtration with $c_1 = 1$ with rescaling factor $c > 1$. However, as we shall see in (12.4), if Δ is stably contact, then its rescaling $\{\Delta(c\nu)\}$ is stably contact provided $c \geq 1$. (Note that a similar statement for the contact filtration is completely obvious from the definitions, since rescaling with the factor $c > 1$ relaxes both conditions (9.2.1) and (9.2.2).)

(12.3) Main Lemma. *Let Δ be a stably contact filtration in $\mathcal{O}$ and u^α a fractional monomial in $\mathcal{O}$. Suppose $\Delta \subset \{\mathcal{M}^\nu\}$ and let $c \in \mathbf{Q}_+$ be such that*

(12.3.1) $|\alpha| + c > 1$

(12.3.2) either $c \geq 1$ *or*

$$u^{\frac{c}{1-c}\cdot\alpha} \cdot (1) \subset \{\Delta(c\nu)\}$$

(cf. (7.6)). Then $u^\alpha \cdot \{\Delta(c\nu)\}$ is a stably contact filtration in $\mathcal{O}$.

This key lemma is very helpful, and we shall use it both here and in some other places later.

(12.4) Corollary. *If Δ is stably contact and $c \geq 1$, then the rescaling $\{\Delta(c\nu)\}$ is also stably contact. In particular, $\{\mathcal{M}^{c\nu}\}$ is stably contact if $c \geq 1$.*

(Indeed, we just have to apply (12.3) with $\alpha = 0$.) ∎

(12.5) Proof of Proposition 12.1: Indeed, we apply the inductive description (7.6) of the special filtrations. We reformulate this inductive description in the following way.

The base of the induction: the three filtrations (0), $\{\mathcal{M}^\nu\}$, and $u^\alpha \cdot (1)$ with $|\alpha| > 1$. (Note that all these three filtrations are contained in $\{\mathcal{M}^\nu\}$.)

The induction step: take Δ from the previous step (which will always mean $\Delta \subset \{\mathcal{M}^\nu\}$), take $c \in \mathbf{Q}_+$, take a fractional monomial u^α such that (12.3.1) and (12.3.2) are satisfied, and take a k-fold suspension ($k \geq 1$) of the filtration $u^\alpha \cdot \{\Delta(c\nu)\}$ — this is the new special filtration. (Note that (12.3.1) guarantees that the result of each step is contained in $\{\mathcal{M}^\nu\}$.)

At the end of the induction process we may rescale the final filtration by a factor $c \geq 1$.

As we have seen in (11.4), all three filtrations which form the base of our inductive construction, are stably contact. (12.3) and (11.5) show that each step also produces a stably contact filtration. Corollary 12.4 shows that the final rescaling also preserves stable contactness. ■

Now we turn to the proof of the Main Lemma 12.3.

(12.6) Lemma. *Let u^α be a fractional monomial in $\mathcal{O}$, and let Δ be an integrally closed filtration in $\mathcal{O}[\sqrt{u}]$. Consider a filtration Δ' in $\mathcal{O}[\sqrt{u}]$ defined by*

$$\Delta'(\nu) = u^{\nu\alpha} \cdot \Delta(\nu)$$

(We shall also use the notation $\Delta' = u^\alpha \cdot \Delta$.) Then Δ' is also integrally closed.

The proof of this lemma is given below in (12.10).

(12.7) Lemma. *Let $\Delta \neq (1)$ be a stably contact filtration in $\mathcal{O}$. Let $\widehat{\Delta}$ be the minimal filtration in $\mathcal{O}[\sqrt{u}]$ which is integrally closed, satisfies (11.2.1), and contains Δ.*

Suppose $c \in \mathbf{Q}_+$ and a fractional monomial u^α in $\mathcal{O}$ satisfy the following property:

(12.7.1) either $c \geq 1$ or

$$u^{\frac{c}{1-c}\cdot\alpha} \cdot (1) \subset \{\Delta(c\nu)\}$$

(cf. (7.6), (12.3.2)).

Consider a filtration $\widehat{\Delta}'$ in $\mathcal{O}[\sqrt{u}]$ defined by

$$\widehat{\Delta}' = u^\alpha \cdot \{\widehat{\Delta}(c\nu)\}$$

(i.e., $\widehat{\Delta}'(\nu) = u^{\nu\alpha} \cdot \widehat{\Delta}(c\nu)$). Then $\widehat{\Delta}'$ satisfies the following analog of the property (11.2.1):

$$\text{(12.7.2)} \qquad \mathrm{Der}_{\mathcal{O},I}\,\widehat{\Delta}'(\nu) \subset \widehat{\Delta}'(\nu - 1)\,.$$

(Recall that, as we have noted in (11.1), $\mathrm{Der}_{\mathcal{O},I}$ acts by everywhere defined derivations on $\mathcal{O}[\sqrt{u}]$, as opposed to $\mathrm{Der}_\mathcal{O}$ which acts by rational derivations.)

Lemma 12.7 is actually the key ingredient in the proof of the Main Lemma 12.3; its proof is given below in (12.11).

(12.8) Lemma. *Let Δ, Δ' be integrally closed filtrations in $\mathcal{O}$ and $\mathcal{O}[\sqrt{u}]$ respectively, such that $\Delta = \Delta' \cap \mathcal{O}$. Let u^α be a fractional monomial in $\mathcal{O}$; then*

$$(u^\alpha \cdot \Delta') \cap \mathcal{O} = u^\alpha \cdot \Delta\,.$$

The proof is given below in (12.12).

(12.9) Proof of the Main Lemma 12.3: Let $\widehat{\Delta}$ be the minimal filtration in $\mathcal{O}[\sqrt{u}]$ which is integrally closed, satisfies (11.2.1), and contains Δ. Note that the condition (11.2.1) involves the module $\mathrm{Der}_{\mathcal{O},I_{\mathrm{tr}}(\Delta)}$ which depends on the original filtration Δ. As we shall consider stable contactness properties of different filtrations, we shall have to specify explicitly which original filtration we refer to in (11.2.1). For example, when we referred to $\widehat{\Delta}$, the requirement was that $\widehat{\Delta}$ satisfied the condition (11.2.1) for Δ.

Now assume that c and u^α satisfy (12.3.1) and (12.3.2), and let $\Delta' = u^\alpha \cdot \{\Delta(c\nu)\}$. Consider the minimal filtration $\widehat{\Delta}'$ in $\mathcal{O}[\sqrt{u}]$ which is integrally closed, satisfies (11.2.1) for Δ', and contains Δ'.

Note that since $\Delta \subset \{\mathcal{M}^{\nu}\}$ and $|\alpha|+c>1$, $\Delta' = u^{\alpha}\cdot\{\Delta(c\nu)\} \subset \{\mathcal{M}^{\nu+0}\}$, so $I_{\mathrm{tr}}(\Delta') = I$, which means that Δ' is transverse to all the fixed variables of $\mathcal{O}$.

We have to prove that Δ' is stably contact; by Remark 11.15 this means that Δ' satisfies the condition (9.2.2) and that $\widehat{\Delta}'\cap\mathcal{O} = \Delta'$. Clearly, in our case (9.2.2) is a vacuous condition, since there are no variables that are intransverse to Δ'; thus, all we have to prove is that $\widehat{\Delta}'\cap\mathcal{O} = \Delta'$.

Note also that since $I_{\mathrm{tr}}(\Delta') = I$, the condition (11.2.1) for Δ' amounts to

$$\mathrm{Diff}_{\mathcal{O},I}\,\widehat{\Delta}'(\nu) \subset \widehat{\Delta}'(\nu-1)\,.$$

Now denote $\widehat{\Delta}'' = u^{\alpha}\cdot\{\widehat{\Delta}(c\nu)\}$ (i.e., $\widehat{\Delta}''(\nu) = u^{\nu\alpha}\cdot\widehat{\Delta}(c\nu)$); by Lemma 12.6 this is an integrally closed filtration in $\mathcal{O}[\sqrt{u}]$. Lemma 12.7 yields $\mathrm{Diff}_{\mathcal{O},I}\,\widehat{\Delta}''(\nu)\subset\widehat{\Delta}''(\nu-1)$, which means that $\widehat{\Delta}''$ also satisfies the condition (11.2.1) for Δ'. As clearly $\widehat{\Delta}''\supset\Delta'$, the minimality of $\widehat{\Delta}'$ yields $\widehat{\Delta}'\subset\widehat{\Delta}''$.

Now let us apply Lemma 12.8 to the filtrations $\{\Delta(c\nu)\}$ and $\{\widehat{\Delta}(c\nu)\}$ in $\mathcal{O}$ and $\mathcal{O}[\sqrt{u}]$ respectively; we know that $\{\widehat{\Delta}(c\nu)\}\cap\mathcal{O} = \{\Delta(c\nu)\}$ (i.e., $\widehat{\Delta}\cap\mathcal{O}=\Delta$) since Δ is stably contact. Lemma 12.8 then yields

$$\left[u^{\alpha}\cdot\{\widehat{\Delta}(c\nu)\}\right]\cap\mathcal{O} = u^{\alpha}\cdot\{\Delta(c\nu)\}$$

i.e.,

$$\widehat{\Delta}''\cap\mathcal{O} = \Delta'\,.$$

Now $\Delta'\subset\widehat{\Delta}'\cap\mathcal{O}\subset\widehat{\Delta}''\cap\mathcal{O}=\Delta'$; thus, $\widehat{\Delta}'\cap\mathcal{O}=\Delta'$, and this proves that Δ' is stably contact. ∎

(12.10) Proof of Lemma 12.6: First of all, note that the statement of the lemma is clear in case $\Delta = (1)$. In the general case it is clear that Δ' satisfies the multiplicativity properties (2.1.1) and (2.1.2), so only the integral closedness (2.1.3) requires proof.

Let $f\in\mathcal{O}[\sqrt{u}]$ satisfy an equation

$$f^n + g_1 f^{n-1} + g_2 f^{n-2} + \ldots + g_n = 0 \tag{12.10.1}$$

with $g_i\in\Delta'(i\nu) = u^{i\nu\alpha}\Delta(i\nu)$. Then the statement of the lemma for the case $\Delta=(1)$ yields $f\in(u^{\nu\alpha})$, i.e., $f = u^{\nu\alpha}\widetilde{f}$, where $\widetilde{f}\in\mathcal{O}[\sqrt{u}]$. The equality (12.10.1) means that $\widetilde{f}$ satisfies an equation

$$\widetilde{f}^n + \widetilde{g}_1\widetilde{f}^{n-1} + \widetilde{g}_2\widetilde{f}^{n-2} + \ldots + \widetilde{g}_n = 0$$

where $\widetilde{g}_i = g_i/u^{i\nu\alpha}\in\Delta(i\nu)$. Thus, integral closedness of Δ yields $\widetilde{f}\in\Delta(\nu)$ and consequently $f\in\Delta'(\nu)$. ∎

(12.11) Proof of Lemma 12.7: Consider the left-hand side of (12.7.2):

$$\begin{aligned}\mathrm{Der}_{\mathcal{O},I}\,\widehat{\Delta}'(\nu) &= \mathrm{Der}_{\mathcal{O},I}(u^{\nu\alpha}\widehat{\Delta}(c\nu))\\ &\subset (\mathrm{Der}_{\mathcal{O},I}\,u^{\nu\alpha})\cdot\widehat{\Delta}(c\nu) + u^{\nu\alpha}\cdot\mathrm{Der}_{\mathcal{O},I}\,\widehat{\Delta}(c\nu)\\ &= u^{\nu\alpha}\cdot\widehat{\Delta}(c\nu) + u^{\nu\alpha}\cdot\mathrm{Der}_{\mathcal{O},I}\,\widehat{\Delta}(c\nu)\,.\end{aligned} \tag{12.11.1}$$

The first term in (12.11.1) is just $u^{\nu\alpha}\cdot\widehat{\Delta}(c\nu) = \widehat{\Delta}'(\nu)$. In the second term

$$\mathrm{Der}_{\mathcal{O},I}\,\widehat{\Delta}(c\nu)\subset\widehat{\Delta}(c\nu-1)$$

since $\widehat{\Delta}$ satisfies the property (11.2.1) and $\mathrm{Der}_{\mathcal{O},I}\subset\mathrm{Der}_{\mathcal{O},I_1(\Delta)}$. If $c\geq 1$, then

$$u^{\nu\alpha}\cdot\mathrm{Der}_{\mathcal{O},I}\,\widehat{\Delta}(c\nu)\subset u^{\nu\alpha}\cdot\widehat{\Delta}(c\nu-1)\subset u^{(\nu-1)\alpha}\widehat{\Delta}(c(\nu-1)) = \widehat{\Delta}'(\nu-1)\,. \tag{12.11.2}$$

If $c < 1$, then it follows from (12.7.1) that

$$u^\alpha \in \widehat{\Delta}\left(c \cdot \frac{1-c}{c}\right) = \widehat{\Delta}(1-c)$$

and

$$\begin{aligned} u^{\nu\alpha} \cdot \mathrm{Der}_{\mathcal{O},I}\, \widehat{\Delta}(c\nu) &\subset u^{(\nu-1)\alpha} \cdot u^\alpha \cdot \widehat{\Delta}(c\nu - 1) \\ &\subset u^{(\nu-1)\alpha} \cdot \widehat{\Delta}(1-c) \cdot \widehat{\Delta}(c\nu - 1) \\ &= u^{(\nu-1)\alpha} \cdot \widehat{\Delta}(c(\nu-1)) = \widehat{\Delta}'(\nu - 1) \ . \end{aligned} \tag{12.11.3}$$

Thus, in any case

$$\begin{aligned} \mathrm{Der}_{\mathcal{O},I}\, \widehat{\Delta}(\nu) &\subset u^{\nu\alpha} \cdot \widehat{\Delta}(c\nu) + u^{\nu\alpha} \cdot \mathrm{Der}_{\mathcal{O},I}\, \widehat{\Delta}(c\nu) \\ &\subset \widehat{\Delta}'(\nu) + \widehat{\Delta}'(\nu-1) = \widehat{\Delta}'(\nu-1) \ . \end{aligned}$$

■

(12.12) Proof of Lemma 12.8: It is easy to see — either from the definition of $u^\alpha \cdot \Delta$ or from (6.9) — that

$$u^\alpha \cdot \Delta \subset (u^\alpha \cdot \Delta') \cap \mathcal{O} \ . \tag{12.12.1}$$

Now let $\Delta'' = u^\alpha \cdot \Delta'$; the same reasoning shows that

$$(\Delta'' \cap \mathcal{O}) : u^\alpha \subset (\Delta'' : u^\alpha) \cap \mathcal{O} \ . \tag{12.12.2}$$

However, $\Delta'' \cap \mathcal{O} = (u^\alpha \cdot \Delta') \cap \mathcal{O}$ and $(\Delta'' : u^\alpha) \cap \mathcal{O} = \Delta' \cap \mathcal{O} = \Delta$; thus, (12.12.2) can be rewritten as

$$\left[(u^\alpha \cdot \Delta') \cap \mathcal{O}\right] : u^\alpha \subset \Delta \ . \tag{12.12.3}$$

However, $(u^\alpha \cdot \Delta') \cap \mathcal{O} \subset u^\alpha \cdot (1)$ (as filtrations in $\mathcal{O}$), and by (6.7.6) and (6.11),

$$(u^\alpha \cdot \Delta') \cap \mathcal{O} = u^\alpha \cdot \left[\left[(u^\alpha \cdot \Delta') \cap \mathcal{O}\right] : u^\alpha\right] \subset u^\alpha \cdot \Delta \ . \tag{12.12.4}$$

Thus, (12.12.1) and (12.12.4) together prove (12.8). ■

(12.13) Remark. It is not clear whether the Main Lemma 12.3 will be true if we substitute "contact" for "stably contact", i.e., whether $u^\alpha \cdot \{\Delta(c\nu)\}$ will be contact if Δ is. Proposition 12.1, of course, will be still true if we substitute "contact" for "stably contact", and we could have proved this without appealing to the notion of stable contactness.

(12.14) Remark. The following corollary to Lemma 12.8 will be useful for us later :

Let Δ , Δ' and u^α be the same as in (12.8); then

$$(\Delta' : u^\alpha) \cap \mathcal{O} = \Delta : u^\alpha \ . \tag{12.14.1}$$

Indeed, applying Lemma 12.8 to $(\Delta' : u^\alpha) \cap \mathcal{O}$ and $\Delta' : u^\alpha$, we get

$$\left(u^\alpha \cdot (\Delta' : u^\alpha)\right) \cap \mathcal{O} = u^\alpha \cdot [(\Delta' : u^\alpha) \cap \mathcal{O}] \ . \tag{12.14.2}$$

However,

$$\left(u^\alpha \cdot (\Delta' : u^\alpha)\right) \cap \mathcal{O} = \left(\Delta' \cap u^\alpha \cdot \widehat{(1)}\right) \cap \mathcal{O} = \Delta \cap u^\alpha \cdot (1) = u^\alpha \cdot (\Delta : u^\alpha) \ . \tag{12.14.3}$$

(Here by $\widehat{(1)}$ we denote the filtration (1) in $\mathcal{O}[\sqrt{u}]$, to distinguish it from the filtration (1) in $\mathcal{O}$.)

Now from (12.14.2) and (12.14.3) we see that

$$u^\alpha \cdot [(\Delta' : u^\alpha) \cap \mathcal{O}] = u^\alpha \cdot (\Delta : u^\alpha)$$

which yields (12.14.1).

Chapter 3. The first derived filtration and its structure

In this chapter we apply the machinery of stably contact filtrations to construct for any filtration Δ another filtration $\mathrm{Dr}_1\,\Delta \supset \Delta$. If Δ is special, then $\mathrm{Dr}_1\,\Delta = \Delta$, and if Δ^* is the special filtration associated to a filtration Δ (the existence and the uniqueness of such Δ^* is the subject of the Main Theorem 7.11), then $\mathrm{Dr}_1\,\Delta \subset \Delta^*$.

The major part of this chapter is devoted to the study of certain finiteness properties of $\mathrm{Dr}_1\,\Delta$.

13. First coweight and the first derived filtration $\mathrm{Dr}_1\,\Delta$

Let $\mathcal{O}$ be a local ring with the maximal ideal $\mathcal{M}$.

(13.1) Definition. *Let Δ be an integrally closed filtration in $\mathcal{O}$. We define its first coweight $\mathrm{cw}_1(\Delta) = \mathrm{cw}_1\,\Delta$ by*

$$\mathrm{cw}_1\,\Delta = \sup\Big\{c \in \mathbf{Q}_+ \;\Big|\; \Delta \subset \{\mathcal{M}^{c\nu}\}\Big\}$$

By definition, the first coweight is a nonnegative real number or ∞.

(13.2) Examples and Remarks. $\mathrm{cw}_1[(1)] = 0$, $\mathrm{cw}_1[(0)] = \infty$, $\mathrm{cw}_1\{\mathcal{M}^{c\nu}\} = c$. If Δ is quasi-homogeneous, then $\mathrm{cw}_1\,\Delta$ is the smallest of its coweights. For a rescaling, $\mathrm{cw}_1\{\Delta(c\nu)\} = c \cdot \mathrm{cw}_1\,\Delta$. If $\Delta \subset \Delta'$, then $\mathrm{cw}_1\,\Delta \geq \mathrm{cw}_1\,\Delta'$.

(13.3) Proposition. *Suppose Δ is generated by a family f_α, $\alpha \in A$, with the weights $\nu_\alpha > 0$. Then*

$$\mathrm{cw}_1\,\Delta = \inf_{\alpha \in A} \frac{\nu(f_\alpha)}{\nu_\alpha}$$

where $\nu(f_\alpha)$ is the integer satisfying $f_\alpha \in \mathcal{M}^{\nu(f_\alpha)} \setminus \mathcal{M}^{\nu(f_\alpha)+1}$.

(13.4) Proof: Let $c \in \mathbf{Q}_+$ be such that $\Delta \subset \{\mathcal{M}^{c\nu}\}$; then $f_\alpha \in \Delta(\nu_\alpha) \subset \mathcal{M}^{c\nu_\alpha}$. Thus, $c\nu_\alpha \leq \nu(f_\alpha)$. Consequently, $c \leq \nu(f_\alpha)/\nu_\alpha$ and $c \leq \inf \nu(f_\alpha)/\nu_\alpha$. Since this is true for any c such that $\Delta \subset \{\mathcal{M}^{c\nu}\}$, we see that $\mathrm{cw}_1\,\Delta \leq \inf \nu(f_\alpha)/\nu_\alpha$.

Now take any $c \in \mathbf{Q}_+$, $c \leq \inf \nu(f_\alpha)/\nu_\alpha$. Then $c\nu_\alpha \leq \nu(f_\alpha)$ for any α, and $f_\alpha \in \mathcal{M}^{\nu(f_\alpha)} \subset \mathcal{M}^{c\nu_\alpha}$. This shows that each generator f_α of Δ lies in ν_α-th term of $\{\mathcal{M}^{c\nu}\}$, so $\Delta \subset \{\mathcal{M}^{c\nu}\}$ and $\mathrm{cw}_1\,\Delta \geq c$. Thus, $\mathrm{cw}_1\,\Delta = \inf \nu(f_\alpha)/\nu_\alpha$. ∎

(13.5) Remark. In Section 4 we defined the filtration generated by a family f_α, $\alpha \in A$, with the weights ν_α, and there we required only that $\nu_\alpha \in \mathbf{Q}_+$, i.e., we allowed the possibility of $\nu_\alpha = 0$. However, it is clear that dropping all the generators with zero weights does not change the filtration, and we shall assume that the generators of any filtration always have positive weights, i.e., we shall always assume $\nu_\alpha > 0$.

(13.6) Corollary to Proposition 13.3.

(13.6.1) $\operatorname{cw}_1(\Delta_1 * \Delta_2) = \min(\operatorname{cw}_1 \Delta_1, \operatorname{cw}_1 \Delta_2)$

(13.6.2) $\operatorname{cw}_1(\{(x_1,\ldots,x_n)^\nu\} * \Delta) = \min(1, \operatorname{cw}_1 \Delta)$

(13.6.3) $\operatorname{cw}_1(u^\alpha \cdot \Delta) = |\alpha| + \operatorname{cw}_1 \Delta$

(13.7) Proof: Indeed, each of the filtrations on the left-hand side is defined as a minimal integrally closed filtration containing a certain family of elements (e.g., $\Delta_1 * \Delta_2$ is generated by all $f \in \Delta_1(\nu) \cup \Delta_2(\nu)$ for all ν). Now we can apply Proposition 13.3, and it yields all the results. ■

(13.8) Corollary to Proposition 13.3. *If Δ is finitely generated, then $\operatorname{cw}_1 \Delta$ is rational. If in addition $\Delta \neq (1)$, then $\operatorname{cw}_1 \Delta > 0$.*

(13.9) Proof: Indeed, if Δ is generated by a finite family f_α , $\alpha \in A$, with the weights $\nu_\alpha > 0$, then

$$\operatorname{cw}_1 \Delta = \inf_{\alpha \in A} \frac{\nu(f_\alpha)}{\nu_\alpha} = \min_{\alpha \in A} \frac{\nu(f_\alpha)}{\nu_\alpha} \in \mathbf{Q}_+ .$$

If $\operatorname{cw}_1 \Delta = 0$, then $\nu(f_\alpha) = 0$ for some α , which means that f_α is invertible and thus $\Delta = (1)$, since $\nu_\alpha > 0$. ■

From now on let the ambient ring $\mathcal{O}$ be a complete normal crossing ring. Let Δ be an integrally closed filtration such that $\operatorname{cw}_1 = \operatorname{cw}_1 \Delta$ is rational.

(13.10) Proposition. *In the situation above, we have $\Delta \subset \{\mathcal{M}^{\operatorname{cw}_1 \cdot \nu}\}$.*

(13.11) Proof: Indeed,

$$\Delta \subset \bigcap_{c < \operatorname{cw}_1} \{\mathcal{M}^{c\nu}\} = \{\mathcal{M}^{\operatorname{cw}_1 \cdot \nu}\} .$$

■

(13.12) Corollary. *If in addition $0 < \operatorname{cw}_1 < \infty$, then $\{\Delta(\nu/\operatorname{cw}_1)\} \subset \{\mathcal{M}^\nu\}$ and there exists a stably contact filtration $\Delta' \neq (1)$ which is minimal among all stably contact filtrations containing $\{\Delta(\nu/\operatorname{cw}_1)\}$.*

■

(13.13) Notation. $I_1(\Delta) = I_{\mathrm{tr}}(\Delta(\nu/\operatorname{cw}_1))$.

(13.14) Definition. *The first derived filtration $\operatorname{Dr}_1 \Delta$ is the "rescaling back" of this filtration Δ' , namely, $\operatorname{Dr}_1 \Delta$ is the filtration determined by the property that its rescaling $\{(\operatorname{Dr}_1 \Delta)(\nu/\operatorname{cw}_1)\}$ is the minimal stably contact filtration containing $\{\Delta(\nu/\operatorname{cw}_1)\}$. In case $\operatorname{cw}_1 = 0$ or $\operatorname{cw}_1 = \infty$ (i.e., $\Delta = (1)$ or $\Delta = (0)$), we shall say that $\operatorname{Dr}_1 \Delta$ is undefined. We shall also say that $\operatorname{Dr}_1 \Delta$ is undefined, if $\operatorname{cw}_1 \Delta$ is not a rational number.*

(13.15) Remark. We may apply the structure results on contact and stably contact filtrations to $\{(\operatorname{Dr}_1 \Delta)(\nu/\operatorname{cw}_1)\}$. This yields the following results on $\operatorname{Dr}_1 \Delta$:

(13.15.1) *Let $\mathcal{O} = \mathcal{O}_1[[x_1, x_2, \ldots, x_{m_1}]]$ as normal crossing rings, and suppose*

$$x_1, x_2, \ldots, x_{m_1} \in (\operatorname{Dr}_1 \Delta)(1/\operatorname{cw}_1)$$

and the images of $x_1, x_2, \ldots, x_{m_1}$ *in* $\mathcal{M}/\mathcal{M}^2$ *form a basis of the subspace* $[(\mathrm{Dr}_1\,\Delta)(1/\mathrm{cw}_1) \bmod \mathcal{M}^2] \subset \mathcal{M}/\mathcal{M}^2$. *Then there exists a unique filtration* $\Delta^{(1)}$ *in* $\mathcal{O}_1$ *such that*

$$\mathrm{Dr}_1\,\Delta = \left\{(x_1, x_2, \ldots, x_{m_1})^{\mathrm{cw}_1 \cdot \nu}\right\} * \Delta^{(1)}$$

In addition, $\{\Delta^{(1)}(\nu/\mathrm{cw}_1)\}$ *is stably contact, and* $\Delta^{(1)} \subset \{\mathcal{M}^{\mathrm{cw}_1 \cdot \nu + 0}\}$.*

If Δ is generated by a family f_α , $\alpha \in A$, with the weights ν_α , then the subspace

$$[(\mathrm{Dr}_1\,\Delta)(1/\mathrm{cw}_1) \bmod \mathcal{M}^2] \subset \mathcal{M}/\mathcal{M}^2$$

can be described in the following ways:

(13.15.2) *For each* α *such that* $\mathrm{cw}_1 \cdot \nu_\alpha \in \mathbf{Z}$, *consider the initial form*

$$[f_\alpha \bmod \mathcal{M}^{\mathrm{cw}_1 \cdot \nu_\alpha + 1}] \in \mathcal{M}^{\mathrm{cw}_1 \cdot \nu_\alpha}/\mathcal{M}^{\mathrm{cw}_1 \cdot \nu_\alpha + 1}$$

(we are using here the fact that $f_\alpha \in \Delta(\nu_\alpha) \subset \mathcal{M}^{\mathrm{cw}_1 \cdot \nu_\alpha}$ *for each* α *). Then* $(\mathrm{Dr}_1\,\Delta)(1/\mathrm{cw}_1) \bmod \mathcal{M}^2$ *is the minimal subspace in* $\mathcal{M}/\mathcal{M}^2$, *such that it is weakly transverse to the fixed variables and all these initial forms lie in the polynomial subalgebra of* $\mathrm{gr}\,\mathcal{O}$ *generated by this subspace.*

(13.15.3) $[(\mathrm{Dr}_1\,\Delta)(1/\mathrm{cw}_1) \bmod \mathcal{M}^2] = [\mathcal{B} \bmod \mathcal{M}^2]$, *where*

$$\mathcal{B} = \sum_{\{\alpha \in A \mid \mathrm{cw}_1 \cdot \nu_\alpha \in \mathbf{Z}_+\}} \mathrm{Diff}_{I_1(\Delta)}^{\mathrm{cw}_1 \cdot \nu_\alpha - 1} f + \sum_{i \in I \setminus I_1(\Delta)} (u_i)$$

(13.16) Definition. *If* $\mathrm{cw}_1\,\Delta$ *is rational,* $0 < \mathrm{cw}_1\,\Delta < \infty$, *then we shall say that*

$$m_1 = \mathrm{m}_1(\Delta) = \mathrm{m}_1\,\Delta = \dim[(\mathrm{Dr}_1\,\Delta)(1/\mathrm{cw}_1\,\Delta) \bmod \mathcal{M}^2]$$

is the multiplicity of the first coweight of Δ .

(13.17) Remark. The following is a direct corollary to the definitions:

(13.17.1) $\mathrm{Dr}_1\,\mathrm{Dr}_1\,\Delta = \mathrm{Dr}_1\,\Delta$

(13.17.2) $\mathrm{cw}_1\,\mathrm{Dr}_1\,\Delta = \mathrm{cw}_1\,\Delta$

(13.17.3) $\mathrm{m}_1\,\mathrm{Dr}_1\,\Delta = \mathrm{m}_1\,\Delta$

(13.17.4) If $\Delta \subset \Delta'$, $\mathrm{cw}_1\,\Delta = \mathrm{cw}_1\,\Delta'$, then $\mathrm{Dr}_1\,\Delta \subset \mathrm{Dr}_1\,\Delta'$, $\mathrm{m}_1\,\Delta \le \mathrm{m}_1\,\Delta'$, $I_1\Delta \supset I_1\Delta'$

(13.17.5) $\mathrm{m}_1\{\Delta(c\nu)\} = \mathrm{m}_1\,\Delta$

(13.17.6) $\mathrm{m}_1\,\Delta \le \dim \mathcal{O}$

(13.17.7) $\mathrm{m}_1\{\mathcal{M}^{c\nu}\} = \dim \mathcal{O}$ and if $\mathrm{m}_1\,\Delta = \dim \mathcal{O}$, then $\mathrm{Dr}_1\,\Delta = \{\mathcal{M}^{c\nu}\}$, where $c = \mathrm{cw}_1\,\Delta$

(13.17.8) $\mathrm{m}_1[(0)] = \mathrm{m}_1\{\mathcal{M}^{c\nu+0}\} = 0$, and $\mathrm{m}_1\,\Delta = 0$ if and only if $\Delta \subset \{\mathcal{M}^{c\nu+0}\}$, where $c = \mathrm{cw}_1\,\Delta$

(13.17.9) If Δ is quasihomogeneous, then $\mathrm{m}_1\,\Delta$ is indeed the multiplicity of the smallest coweight.

* $\{\mathcal{M}^{c\nu+0}\}$ is the rescaling of $\{\mathcal{M}^{\nu+0}\}$; it is the subfiltration of $\{\mathcal{M}^{c\nu}\}$ containing those elements whose (appropriately defined) initial forms are zero.

(13.18) Proposition. *If* Δ *is finitely generated,* $\Delta \neq (0),(1)$ *, then* $\mathrm{m}_1 \Delta > 0$.

(13.19) Remark. If $\Delta \neq (0),(1)$ is finitely generated, then both $\mathrm{Dr}_1 \Delta$ and $\mathrm{m}_1 \Delta$ are defined, since $\mathrm{cw}_1 \Delta$ is rational by (13.8).

(13.20) Proof of (13.18): $\mathrm{m}_1 \Delta = \dim[(\mathrm{Dr}_1 \Delta)(1/\mathrm{cw}_1) \bmod \mathcal{M}^2]$, where $\mathrm{cw}_1 = \mathrm{cw}_1 \Delta$, and we apply the description of the subspace $[(\mathrm{Dr}_1 \Delta)(1/\mathrm{cw}_1) \bmod \mathcal{M}^2] \subset \mathcal{M}/\mathcal{M}^2$ given in (13.15.2). It shows that $\mathrm{m}_1 \Delta = 0$ is possible only if for each generator f_α either $\mathrm{cw}_1 \cdot \nu_\alpha \notin \mathbf{Z}$ or $f_\alpha \in \mathcal{M}^{\mathrm{cw}_1 \cdot \nu_\alpha + 1}$. However, this means that in any case $\nu(f_\alpha) > \mathrm{cw}_1 \cdot \nu_\alpha$ for all α . This, however, contradicts (13.3) if the number of generators is finite. ∎

(13.21) Proposition. *If* Δ *is a special filtration, then* $\mathrm{Dr}_1 \Delta = \Delta$.

(13.22) Remark. The inductive construction of the special filtrations shows that they are finitely generated, so $\mathrm{Dr}_1 \Delta$ and $\mathrm{m}_1 \Delta$ are defined if Δ is special.

(13.23) Proof of (13.21): Indeed, let Δ be a special filtration, $\mathrm{cw}_1 = \mathrm{cw}_1 \Delta$. Then $\{\Delta(\nu/\mathrm{cw}_1)\}$ is also a special filtration, and $\{\Delta(\nu/\mathrm{cw}_1)\} \subset \{\mathcal{M}^\nu\}$. By (12.1), $\{\Delta(\nu/\mathrm{cw}_1)\}$ is stably contact, so $\{\mathrm{Dr}_1 \Delta(\nu/\mathrm{cw}_1)\} = \{\Delta(\nu/\mathrm{cw}_1)\}$ and $\mathrm{Dr}_1 \Delta = \Delta$. ∎

(13.24) Remark. If Δ is special, then its first characteristic string has the form $(\mathrm{cw}_1 \Delta, -\mathrm{m}_1 \Delta, \ldots)$, where $\ldots$ stands for other terms.

(13.25) Corollary. *If* $\Delta \subset \Delta'$ *are two integrally closed filtrations for which* $\mathrm{Dr}_1 \Delta$ *and* $\mathrm{Dr}_1 \Delta'$ *are defined, then* $(\mathrm{cw}_1 \Delta, -\mathrm{m}_1 \Delta) \geq (\mathrm{cw}_1 \Delta', -\mathrm{m}_1 \Delta')$ *where the inequality is understood lexicographically, and in case of equality* $\mathrm{Dr}_1 \Delta \subset \mathrm{Dr}_1 \Delta'$.

(13.26) Proof: Indeed, we have seen that $\mathrm{cw}_1 \Delta \geq \mathrm{cw}_1 \Delta'$, and if $\mathrm{cw}_1 \Delta = \mathrm{cw}_1 \Delta'$, then $\mathrm{Dr}_1 \Delta \subset \mathrm{Dr}_1 \Delta'$ and $\mathrm{m}_1 \Delta \leq \mathrm{m}_1 \Delta'$, i.e., $-\mathrm{m}_1 \Delta \geq -\mathrm{m}_1 \Delta'$. ∎

14. Finitely Der-generated filtrations

(14.1) Motivation. The approach we shall take to the construction of the special filtration Δ^* associated to an integrally closed filtration Δ is as follows. The first step is to take

$$\mathrm{Dr}_1 \Delta = \left\{(x_1, x_2, \ldots, x_{m_1})^{\mathrm{cw}_1 \cdot \nu}\right\} * \Delta^{(1)}$$

(cf. (13.15.1)); if $m_1 < \dim \mathcal{O}$, i.e., $\mathrm{Dr}_1 \Delta \neq \{\mathcal{M}^{\mathrm{cw}_1 \cdot \nu}\}$, then find the maximal fractional monomial u^{mon} such that $\Delta^{(1)} \subset u^{\mathrm{mon}} \cdot (1)$. Then $\Delta^{(1)} = u^{\mathrm{mon}} \cdot \widetilde{\Delta}$ (the details are in Section 16), and

$$\mathrm{Dr}_1 \Delta = \left\{(x_1, x_2, \ldots, x_{m_1})^{\mathrm{cw}_1 \cdot \nu}\right\} * (u^{\mathrm{mon}} \cdot \widetilde{\Delta}) \tag{14.1.1}$$

Comparing this with the inductive description of special filtrations (7.6), we see that special filtrations containing $\widetilde{\Delta}$ correspond to special filtrations containing Δ —if Δ' is a special filtration in $\mathcal{O}_1$ containing $\widetilde{\Delta}$, then it corresponds to the special filtration

$$\Delta'' = \left\{(x_1, x_2, \ldots, x_{m_1})^{\mathrm{cw}_1 \cdot \nu}\right\} * (u^{\mathrm{mon}} \cdot \Delta') \supset \mathrm{Dr}_1 \Delta \supset \Delta \tag{14.1.2}$$

in $\mathcal{O}$ (provided, of course, that the conditions (7.6.1) and (7.6.2) are taken care of; we shall return to this question in Section 17). This suggests the following inductive process for the construction of Δ^* : start with

Δ , construct $\widetilde{\Delta}$ as in (14.1.1) (thus, $\widetilde{\Delta}$ is a filtration in a smaller ring $\mathcal{O}_1 \subset \mathcal{O}$), then take $\mathrm{Dr}_1 \widetilde{\Delta}$ and apply the same procedure to it, and so on. As the dimensions of the ambient rings decrease, this process cannot go on indefinitely, and after some finite number of steps we shall come to one of the three filtrations (0) , (1) , and $\{\mathcal{M}^{c\nu}\}$.

Now we see that the special filtrations containing Δ correspond to the special filtrations containing the final filtration (0) , (1) , or $\{\mathcal{M}^{c\nu}\}$. However, this final filtration is already special, so it is the minimal special filtration containing itself. Thus, it corresponds to some special filtration containing Δ , which is minimal among all the special filtrations containing Δ ; this special filtration can be taken for the filtration Δ^* of the Main Theorem 7.11.

The details of this process will be developed later, but what we see from this brief discussion now is that we need to consider $\mathrm{Dr}_1 \widetilde{\Delta}$, which exists only in case $\mathrm{cw}_1 \widetilde{\Delta}$ is rational and positive. We also need to know that $\mathrm{m}_1 \widetilde{\Delta} > 0$, since otherwise our inductive process will not move forward.

As of now, we have only one way to show that $\mathrm{cw}_1 \widetilde{\Delta}$ is rational and $\mathrm{m}_1 \widetilde{\Delta}$ is positive — we have to show that $\widetilde{\Delta}$ is finitely generated, and then apply (13.8) and (13.18).

However, I do not know whether in general $\Delta^{(1)}$, $\widetilde{\Delta}$, and even $\mathrm{Dr}_1 \Delta$ are finitely generated, if Δ is. To overcome this difficulty, we introduce the following weaker notion of finite generatedness.

(14.2) Definition. *We shall say that an integrally closed filtration Δ in a complete normal crossing ring $\mathcal{O}$ is finitely* Der*-generated, if there exists a finitely generated integrally closed filtration Δ' , such that $\Delta' \subset \Delta$, $\mathrm{cw}_1 \Delta' = \mathrm{cw}_1 \Delta$, and $\mathrm{Dr}_1 \Delta' = \mathrm{Dr}_1 \Delta$. We shall also say that the filtrations (0) and (1) are finitely* Der*-generated.*

(14.3) Remark. Since $\mathrm{cw}_1 \Delta = \mathrm{cw}_1 \Delta'$, it is rational and positive, so $\mathrm{Dr}_1 \Delta$ is defined.

(14.4) Proposition. *Suppose Δ is finitely* Der*-generated. Then*

(14.4.1) $\mathrm{cw}_1 \Delta$ is rational;

(14.4.2) $\mathrm{cw}_1 \Delta = 0$ if and only if $\Delta = (1)$;

(14.4.3) If $\Delta \neq (0), (1)$, then $\mathrm{m}_1 \Delta > 0$;

(14.4.4) There exists some $f \in \mathcal{O}$ such that $f \in \Delta(\nu)$ for some $\nu \in \mathbf{Q}_+$ and $\nu(f) = \nu \cdot \mathrm{cw}_1 \Delta$.

(14.5) Proof: Take a finitely generated filtration $\Delta' \subset \Delta$ such that $\mathrm{Dr}_1 \Delta' = \mathrm{Dr}_1 \Delta$, and apply (13.8), (13.18), and (13.3) to it — this immediately yields the required results about Δ . ■

(14.6) Remark. We see that the finitely Der-generated filtrations have all the good propeties that finitely generated filtrations have. At the same time we can prove finite Der-generatedness of filtrations whose finite generatedness we cannot prove. An example is $\mathrm{Dr}_1 \Delta$, which is clearly finitely Der-generated if Δ is finitely generated.

In Section 16 we shall modify the filtration $\widetilde{\Delta}$ of (14.1) to comply with the properties (7.6.1) and (7.6.2), and show that the resulting filtration $\Delta^{(1+)}$ is finitely Der-generated if Δ is finitely Der-generated.

15. Structure of $\operatorname{Dr}_1 \Delta$ — I. Properties of $\Delta^{(1)}$.

(15.1). In this section Δ is a finitely Der-generated filtration in a complete normal crossing ring $\mathcal{O}$ with the maximal ideal $\mathcal{M}$; we shall assume $\Delta \neq (0),(1)$. As in (13.15.1), let $\mathcal{O} = \mathcal{O}_1[[x_1, x_2, \ldots, x_{m_1}]]$ as normal crossing rings, and assume that all x_i lie in $\operatorname{Dr}_1 \Delta(1/\operatorname{cw}_1)$ and their initial forms $x_i \bmod \mathcal{M}^2$ form a basis of the subspace $[(\operatorname{Dr}_1 \Delta)(1/\operatorname{cw}_1) \bmod \mathcal{M}^2] \subset \mathcal{M}/\mathcal{M}^2$. Let $\mathcal{M}_1$ be the maximal ideal of $\mathcal{O}_1$. Then

$$\operatorname{Dr}_1 \Delta = \{(x_1, x_2, \ldots, x_{m_1})^{\operatorname{cw}_1 \cdot \nu}\} * \Delta^{(1)} \tag{15.1.1}$$

where $\Delta^{(1)}$ is a filtration in $\mathcal{O}_1$. In addition, $\Delta^{(1)} \subset \{\mathcal{M}_1^{\operatorname{cw}_1 \cdot \nu + 0}\}$ and $\{\Delta^{(1)}(\nu/\operatorname{cw}_1)\}$ is stably contact.

(15.2) Proposition. *There exists a finitely generated filtration Δ' in $\mathcal{O}_1$, $\Delta' \subset \Delta^{(1)}$, such that $\{\Delta^{(1)}(\nu/\operatorname{cw}_1)\}$ is the minimal stably contact filtration containing $\{\Delta'(\nu/\operatorname{cw}_1)\}$.*

(15.3) Remark. Clearly, $\operatorname{cw}_1 = \operatorname{cw}_1 \Delta \leq \operatorname{cw}_1 \Delta^{(1)}$ since $\Delta \supset \Delta^{(1)}$. If $\operatorname{cw}_1 \Delta^{(1)} = \operatorname{cw}_1 \Delta = \operatorname{cw}_1$, then the statement of (15.2) would mean that $\Delta^{(1)} = \operatorname{Dr}_1 \Delta'$, i.e., that $\Delta^{(1)}$ is finitely Der-generated. However, as we shall see later in (15.5), this is never the case under our assumptions: the finite Der-generatedness of Δ yields $\operatorname{cw}_1 \Delta < \operatorname{cw}_1 \Delta^{(1)}$.

(15.4) Proof of (15.2): First of all, nothing changes if we shrink the original filtration Δ in such a way that $\operatorname{Dr}_1 \Delta$ is unchanged; thus, we may assume Δ is finitely generated.

Let Δ be generated by $f_1, f_2, \ldots, f_n$ with the weights $\nu_1, \nu_2, \ldots, \nu_n$, and let

$$f_i = \sum_{\alpha=(\alpha_1,\alpha_2,\ldots,\alpha_{m_1}) \in \mathbf{Z}_+^{m_1}} f_{i\alpha}\, x_1^{\alpha_1} x_2^{\alpha_2} \ldots x_{m_1}^{\alpha_{m_1}} \tag{15.4.1}$$

where $f_{i\alpha} \in \mathcal{O}_1$. From (15.1.1) we see that

$$f_{i\alpha} \in \Delta^{(1)}\left(\nu_i - \frac{|\alpha|}{\operatorname{cw}_1}\right). \tag{15.4.2}$$

Let Δ' be the integrally closed filtration in $\mathcal{O}_1$ generated by all $f_{i\alpha}$ (for all i and α) with the weights $\nu_i - \frac{|\alpha|}{\operatorname{cw}_1}$. Clearly, Δ' is actually generated only by those $f_{i\alpha}$ whose weights are positive, i.e., $|\alpha| < \nu_i \cdot \operatorname{cw}_1$. Thus, Δ' is generated by only finitely many $f_{i\alpha}$, so it is finitely generated.

From (15.4.2) we see that $\Delta' \subset \Delta^{(1)}$. We know that $\{\Delta^{(1)}(\nu/\operatorname{cw}_1)\}$ is stably contact; thus, to prove (15.2), it is enough to show that $\{\Delta^{(1)}(\nu/\operatorname{cw}_1)\}$ is the minimal stably contact filtration containing $\{\Delta'(\nu/\operatorname{cw}_1)\}$.

Indeed, let $\Delta'' \supset \{\Delta'(\nu/\operatorname{cw}_1)\}$ be a stably contact filtration in $\mathcal{O}_1$. Consider the filtration

$$\Delta''' = \{(x_1, x_2, \ldots, x_{m_1})^{\nu}\} * \Delta'' \tag{15.4.3}$$

in $\mathcal{O}$; by (11.5) it is also stably contact.

Clearly, $f_{i\alpha} \in \Delta'(\nu_i - \frac{|\alpha|}{\operatorname{cw}_1}) \subset \Delta''(\nu_i \cdot \operatorname{cw}_1 - |\alpha|)$ and consequently $f_i \in \Delta'''(\nu_i \cdot \operatorname{cw}_1)$. Thus, $\Delta \subset \{\Delta'''(\nu \cdot \operatorname{cw}_1)\}$ or equivalently $\{\Delta(\nu/\operatorname{cw}_1)\} \subset \Delta'''$. Since Δ''' is stably contact and $\{(\operatorname{Dr}_1 \Delta)(\nu/\operatorname{cw}_1)\}$ is the minimal stably contact filtration containing $\{\Delta(\nu/\operatorname{cw}_1)\}$, we see that

$$\{(\operatorname{Dr}_1 \Delta)(\nu/\operatorname{cw}_1)\} \subset \Delta'''. \tag{15.4.4}$$

However, from (15.1.1),

$$\{(\mathrm{Dr}_1 \Delta)(\nu/\mathrm{cw}_1)\} = \{(x_1, x_2, \ldots, x_{m_1})^\nu\} * \{\Delta^{(1)}(\nu/\mathrm{cw}_1)\} . \tag{15.4.5}$$

Comparing (15.4.3) with (15.4.5), we find that $\Delta'' \supset \{\Delta^{(1)}(\nu/\mathrm{cw}_1)\}$, and this indeed means that $\{\Delta^{(1)}(\nu/\mathrm{cw}_1)\}$ is the minimal stably contact filtration containing $\{\Delta'(\nu/\mathrm{cw}_1)\}$. ∎

Now we can prove

(15.5) Proposition. $\mathrm{cw}_1 \Delta^{(1)} > \mathrm{cw}_1 \Delta$.

(15.6) Proof: We have already noted in (15.3) that $\mathrm{cw}_1 \Delta^{(1)} \geq \mathrm{cw}_1 \Delta$ since $\Delta^{(1)} \subset \Delta$, so we have to prove that $\mathrm{cw}_1 \Delta^{(1)} \neq \mathrm{cw}_1 \Delta$.

Suppose $\mathrm{cw}_1 \Delta^{(1)} = \mathrm{cw}_1 \Delta = \mathrm{cw}_1$. Then Proposition 15.2 shows that $\Delta^{(1)} = \mathrm{Dr}_1 \Delta'$, where Δ' is a finitely generated filtration in $\mathcal{O}_1$. This yields $\mathrm{m}_1 \Delta^{(1)} = \mathrm{m}_1 \Delta' > 0$, so $[\Delta^{(1)}(1/\mathrm{cw}_1) \bmod \mathcal{M}_1^2] \neq 0$.

However, we know that $\Delta^{(1)} \subset \{\mathcal{M}_1^{\mathrm{cw}_1 \cdot \nu + 0}\}$, which, in particular, means that $[\Delta^{(1)}(1/\mathrm{cw}_1) \bmod \mathcal{M}_1^2] = 0$. This contradiction shows that our assumption $\mathrm{cw}_1 \Delta^{(1)} = \mathrm{cw}_1 \Delta$ is impossible. ∎

Finally, we want to prove the following addition to Proposition 15.2:

(15.7) Proposition. *In the notation of (15.2),* $\mathrm{cw}_1 \Delta' = \mathrm{cw}_1 \Delta^{(1)}$.

(15.8) Proof: Clearly, $\mathrm{cw}_1 \Delta' \geq \mathrm{cw}_1 \Delta^{(1)}$ as $\Delta' \subset \Delta^{(1)}$, and we have to prove the opposite inequality. Let $c = \mathrm{cw}_1 \Delta'/\mathrm{cw}_1$; then $c \geq \mathrm{cw}_1 \Delta^{(1)}/\mathrm{cw}_1$, and by (15.5), $\mathrm{cw}_1 \Delta^{(1)}/\mathrm{cw}_1 > 1$; thus, $c > 1$.

By (12.4), $\{\mathcal{M}^{c\nu}\}$ is stably contact. As $\Delta' \subset \{\mathcal{M}^{\mathrm{cw}_1 \Delta' \cdot \nu}\}$, we have $\{\Delta'(\nu/\mathrm{cw}_1)\} \subset \{\mathcal{M}^{c\nu}\}$; thus,

$$\{\Delta^{(1)}(\nu/\mathrm{cw}_1)\} \subset \{\mathcal{M}^{c\nu}\} \tag{15.8.1}$$

as $\{\Delta^{(1)}(\nu/\mathrm{cw}_1)\}$ is the minimal stably contact filtration containing $\{\Delta'(\nu/\mathrm{cw}_1)\}$.

Now (15.8.1) yields

$$\mathrm{cw}_1\{\Delta^{(1)}(\nu/\mathrm{cw}_1)\} \geq c$$

or equivalently,

$$\mathrm{cw}_1 \Delta^{(1)}/\mathrm{cw}_1 \geq \mathrm{cw}_1 \Delta'/\mathrm{cw}_1$$

or

$$\mathrm{cw}_1 \Delta^{(1)} \geq \mathrm{cw}_1 \Delta' .$$

∎

Having understood these properties of $\Delta^{(1)}$, we may continue our program and turn to the construction of the maximal monomial u^{mon} such that $\Delta^{(1)} = u^{\mathrm{mon}} \cdot \widetilde{\Delta}$.

16. Structure of $\mathrm{Dr}_1\Delta$ — II. The monomial

Here we continue the program outlined in (14.1) and construct the maximal fractional monomial dividing the filtration $\Delta^{(1)}$.

(16.1) Proposition. *Let Δ be any finitely generated integrally closed filtration in a normal crossing ring $\mathcal{O}$. Then there exists a maximal fractional monomial u^{mon} with the property $\Delta \subset u^{\mathrm{mon}} \cdot (1)$; any other monomial with the same property divides u^{mon}.*

We shall say u^{mon} is *the maximal monomial dividing* Δ. In case $\Delta = (0)$ we shall write $\mathrm{mon} = \infty$, meaning that $\mathrm{mon}(i) = \infty$ for each fixed variable u_i of $\mathcal{O}$.

(16.2) Proof: Indeed, let Δ be generated by $f_1, f_2, \ldots, f_n$ with the weights $\nu_1, \nu_2, \ldots, \nu_n$ and let u^{α_i} be the maximal (integer, or "genuine") monomial dividing f_i (such maximal monomial exists since $\mathcal{O}$ is a regular local ring, so it is a unique factorization domain). Let

$$u^{\mathrm{mon}} = \operatorname*{GCD}_{1 \le i \le n}(u^{\alpha_i/\nu_i}) \tag{16.2.1}$$

Then clearly $f_i \in \left[u^{\mathrm{mon}} \cdot (1)\right](\nu_i)$ and thus $\Delta \subset u^{\mathrm{mon}} \cdot (1)$. Clearly, u^{mon} is maximal with the property $\Delta \subset u^{\mathrm{mon}} \cdot (1)$. ■

(16.3) Remark. Finite generatedness is essential in the proof of (16.1), and an attempt to carry out the same proof for infinitely generated Δ may lead to an irrational value of mon in (16.2.1).

(16.4) Proposition. *If Δ is any filtration in a complete local ring $\mathcal{O}$ and u^{mon} is such a fractional monomial that $\Delta \subset u^{\mathrm{mon}} \cdot (1)$, then $\Delta = u^{\mathrm{mon}} \cdot \Delta'$ where $\Delta' = \Delta : u^{\mathrm{mon}}$. (We shall say u^{mon} divides Δ.) In particular, this is true if u^{mon} is the maximal fractional monomial dividing Δ, $\Delta \neq (0)$.*

(The proof is obvious from (6.11).) ■

(16.5). From now on let the notation and the assumptions be the same as in (15.1). In addition, we shall assume $m_1 < \dim\mathcal{O}$ (thus, $\dim\mathcal{O}_1 > 0$). Our aim is to define a fractional monomial u^{mon_1} in $\mathcal{O}_1$ which is maximal such that $\Delta^{(1)} \subset u^{\mathrm{mon}_1} \cdot (1)$. Unfortunately, Proposition 16.1 is not helpful here because $\Delta^{(1)}$ is not necessarily finitely generated, so the existence of such u^{mon_1} still requires proof.

Let the fixed variables of $\mathcal{O}$ be u_i, $i \in I$, and those of $\mathcal{O}_1$ be u_i, $i \in I_1$, where $I_1 = I_1(\Delta) \subset I$.

(16.6) Proposition. *There exists a maximal fractional monomial u^{mon_1} in $\mathcal{O}_1$ such that $\Delta^{(1)} \subset u^{\mathrm{mon}_1} \cdot (1)$. (Again, $\mathrm{mon}_1 = \infty$ if $\Delta^{(1)} = (0)$.)*

(16.7) Lemma. *Let Δ be an integrally closed filtration in a complete normal crossing ring $\mathcal{O}$ with a maximal ideal $\mathcal{M}$ and let Δ' be the minimal stably contact filtration containing it. Assume $\Delta \subset \{\mathcal{M}^{\nu+0}\}$ (this, of course, means that $\Delta' \subset \{\mathcal{M}^{\nu+0}\}$). Then any fractional monomial u^{α} divides Δ' if and only if it divides Δ.*

The proof is given below in (16.11).

(16.8) Remark. The condition $\Delta \subset \{\mathcal{M}^{\nu+0}\}$ is essential for Lemma 16.7; it is easy to see that Lemma 16.7 is not true otherwise.

(16.9) Corollary to Lemma 16.7. *In the notation of (16.7) there exists a maximal monomial dividing Δ' if and only if such monomial exists for Δ, and if these maximal dividing monomials exist for Δ and Δ', then they are equal.*

■

(16.10) Proof of Proposition 16.6: By Proposition 15.2 there exists a finitely generated filtration $\Delta' \subset \Delta^{(1)}$ such that $\{\Delta^{(1)}(\nu/\mathrm{cw}_1)\}$ is the minimal stably contact filtration in $\mathcal{O}_1$ containing $\{\Delta'(\nu/\mathrm{cw}_1)\}$. By (16.1) there exists a maximal monomial u^{mon_1} dividing Δ'; clearly, $u^{\mathrm{mon}_1/\mathrm{cw}_1}$ is the maximal monomial dividing $\{\Delta'(\nu/\mathrm{cw}_1)\}$. By (16.9) $u^{\mathrm{mon}_1/\mathrm{cw}_1}$ is the maximal monomial dividing $\{\Delta^{(1)}(\nu/\mathrm{cw}_1)\}$; thus, u^{mon_1} is the maximal monomial dividing $\Delta^{(1)}$. ■

(16.11) Proof of Lemma 16.7: Clearly, since $\Delta \subset \Delta'$, any monomial dividing Δ' also divides Δ, so we have to prove the opposite statement.

Assume u^α divides Δ; we have to prove that u^α divides Δ'. Let Δ'' be the minimal contact filtration containing Δ; then $\Delta \subset \Delta'' \subset \Delta'$. Let us first show that u^α divides Δ''.

The idea behind the proof is very simple. Clearly, $\Delta \subset \Delta'' \subset \Delta' \subset \{\mathcal{M}^{\nu+0}\}$ and consequently $I_{\mathrm{tr}}(\Delta) = I_{\mathrm{tr}}(\Delta'') = I_{\mathrm{tr}}(\Delta') = I$, i.e., there are no intransverse variables. Thus, the contact closure Δ'' of Δ is more or less the closure of Δ under the action of $\mathrm{Der}_{I_{\mathrm{tr}}(\Delta'')}$. However, since there are no intransverse variables, $\mathrm{Der}_{I_{\mathrm{tr}}(\Delta'')} = \mathrm{Der}_I$ and Der_I preserves divisibility by all monomials.

More precisely, the divisibility condition $\Delta \subset u^\alpha \cdot (1)$ is equivalent to $R_t(\Delta) \subset R_t(u^\alpha \cdot (1))$ where $R_t(\Delta)$ and $R_t(u^\alpha \cdot (1))$ are the corresponding subalgebras in $\mathcal{O}[t^{\mathbf{Q}}]$. Thus, we have to show that if $R_t(\Delta) \subset R_t(u^\alpha \cdot (1))$, then $R_t(\Delta'') \subset R_t(u^\alpha \cdot (1))$.

From (9.3) we see that the subring $R_t(\Delta'')$ can be obtained from $R_t(\Delta)$ by

(a) adding the elements $tu_i \in \mathcal{O}[t^{\mathbf{Q}}]$ for all fixed variables u_i that are intransverse to Δ'', and

(b) taking the closure under the action of $\frac{1}{t}\mathrm{Der}_{I_{\mathrm{tr}}(\Delta'')}$.

Clearly, step (a) is vacuous since there are no intransverse variables. As for step (b), it consists of taking the closure under the action of $\frac{1}{t}\mathrm{Der}_I$, since $I_{\mathrm{tr}}(\Delta'') = I$. Both $\frac{1}{t}$ and Der_I (and consequently $\frac{1}{t}\mathrm{Der}_I$) preserve the subring

$$R_t(u^\alpha \cdot (1)) = \sum_{\nu \in \mathbf{Q}} (u^{\nu\alpha}) \cdot t^\nu .$$

(Here we understand $(u^{\nu\alpha}) = [u^\alpha \cdot (1)](\nu)$ for $\nu\alpha$ fractional.) Thus, $R_t(\Delta'')$, which is the closure of $R_t(\Delta) \subset R_t(u^\alpha \cdot (1))$ under $\frac{1}{t}\mathrm{Der}_I$, also lies in $R_t(u^\alpha \cdot (1))$. This means that $\Delta'' \subset u^\alpha \cdot (1)$ and Δ'' is divisible by u^α.

The proof that the stably contact closure Δ' is also divisible by u^α, goes along the same lines, and it is left to the reader. ■

17. Structure of $\mathrm{Der}_1\Delta$ **— III.** $\Delta^{(1+)}$ **.**

(17.1). Here we carry out the program outlined in (14.1) and construct a filtration $\Delta^{(1+)}$ in a smaller ring $\mathcal{O}_1$ such that special filtrations containing $\Delta^{(1+)}$ correspond to some special filtrations containing the original filtration Δ ; the exact formulation of this correspondence will be given in Section 20.

Let the notation and the assumptions be the same as in (15.1) and (16.5); assume also that $\Delta^{(1)} \neq (0)$.

Let u^{mon_1} be the maximal monomial in $\mathcal{O}_1$ dividing $\Delta^{(1)}$ (it exists by (16.6)), and let $\widetilde{\Delta} = \Delta^{(1)} : u^{\mathrm{mon}_1}$; then by (16.4), $\Delta^{(1)} = u^{\mathrm{mon}_1} \cdot \widetilde{\Delta}$. Note that $\mathrm{mon}_1 \neq \infty$ as $\Delta^{(1)} \neq (0)$; in this case, $\mathrm{mon}_1(i) \neq \infty$ for all i .

Denote $\mathrm{cw}_2 = \mathrm{cw}_1 \widetilde{\Delta}$.

(17.2) Remark. By (13.6.3) $\mathrm{cw}_1 \Delta^{(1)} = |\,\mathrm{mon}_1\,| + \mathrm{cw}_1 \widetilde{\Delta} = |\,\mathrm{mon}_1\,| + \mathrm{cw}_2$. By (15.5), $\mathrm{cw}_1 \Delta^{(1)} > \mathrm{cw}_1 \Delta = \mathrm{cw}_1$; thus, $|\,\mathrm{mon}_1\,| + \mathrm{cw}_2 > \mathrm{cw}_1$ (cf. (7.6.1)).

(17.3) Definition. *We define an integrally closed filtration* $\Delta^{(1+)}$ *in* $\mathcal{O}_1$ *as follows:*

$$\Delta^{(1+)} = \begin{cases} \widetilde{\Delta}\,, & \text{if } \mathrm{cw}_2 \geq \mathrm{cw}_1 \\ \widetilde{\Delta} * \left[u^{\frac{\mathrm{cw}_2}{\mathrm{cw}_1 - \mathrm{cw}_2}\cdot \mathrm{mon}_1} \cdot (1)\right]\,, & \text{if } \mathrm{cw}_2 < \mathrm{cw}_1 . \end{cases}$$

(17.4) Corollary. $\mathrm{cw}_1 \Delta^{(1+)} = \mathrm{cw}_2$.

(17.5) Proof: This is trivial in case $\mathrm{cw}_2 \geq \mathrm{cw}_1$. If $\mathrm{cw}_2 < \mathrm{cw}_1$, then

$$\begin{aligned} \mathrm{cw}_1 \Delta^{(1+)} &= \min\left(\mathrm{cw}_1 \widetilde{\Delta}, \mathrm{cw}_1 \left[u^{\frac{\mathrm{cw}_2}{\mathrm{cw}_1 - \mathrm{cw}_2}\cdot \mathrm{mon}_1} \cdot (1)\right]\right) \\ &= \min\left(\mathrm{cw}_2, \frac{\mathrm{cw}_2}{\mathrm{cw}_1 - \mathrm{cw}_2} \cdot |\,\mathrm{mon}_1\,|\right) . \end{aligned}$$

From $|\,\mathrm{mon}_1\,| + \mathrm{cw}_2 > \mathrm{cw}_1$ we see that $|\,\mathrm{mon}_1\,| > \mathrm{cw}_1 - \mathrm{cw}_2$ and consequently

$$\frac{\mathrm{cw}_2}{\mathrm{cw}_1 - \mathrm{cw}_2} \cdot |\,\mathrm{mon}_1\,| > \mathrm{cw}_2$$

thus, $\mathrm{cw}_1 \Delta^{(1+)} = \mathrm{cw}_2$. ■

(17.6) Explanation. The meaning of Definition 17.3 is as follows. Let Δ' be a special filtration in $\mathcal{O}_1$ containing $\Delta^{(1+)}$, such that $\mathrm{cw}_1 \Delta' = \mathrm{cw}_1 \Delta^{(1+)} = \mathrm{cw}_2$. Consider the filtration Δ'' in $\mathcal{O}$ given by

$$\Delta'' = \left\{(x_1, x_2, \ldots, x_{m_1})^{\mathrm{cw}_1 \cdot \nu}\right\} * (u^{\mathrm{mon}_1} \cdot \Delta') .$$

Then $\Delta'' \supset \Delta$ and Δ'' is an almost special filtration. It is special if the conditions (7.6.1) and (7.6.2) are satisfied, and it is easy to see that the assumption $\mathrm{cw}_1 \Delta' = \mathrm{cw}_2$ guarantees that (7.6.1) is satisfied. As for (7.6.2), it follows from $\Delta' \supset \Delta^{(1+)}$ and the way we defined $\Delta^{(1+)}$ in (17.3) (see the details in (20.8)). The purpose of Definition 17.3 was to make sure that (7.6.2) is satisfied if $\Delta' \supset \Delta^{(1+)}$.

(17.7) Proposition. $\Delta^{(1+)}$ *is finitely* Der*-generated.*

(17.8) Proof: The idea here is very simple. By (15.2) there exists a finitely generated filtration $\Delta' \subset \Delta^{(1)}$ such that $\{\Delta^{(1)}(\nu/\mathrm{cw}_1)\}$ is the minimal stably contact filtration containing $\{\Delta'(\nu/\mathrm{cw}_1)\}$. We shall construct the generators of $\Delta^{(1+)}$ from the generators of Δ' . More precisely, let

$$\Delta'' = \begin{cases} \Delta' : u^{\mathrm{mon}_1}\,, & \text{if } \mathrm{cw}_2 \geq \mathrm{cw}_1 \\ (\Delta' : u^{\mathrm{mon}_1}) * \left[u^{\frac{\mathrm{cw}_2}{\mathrm{cw}_1 - \mathrm{cw}_2}\cdot \mathrm{mon}_1} \cdot (1)\right]\,, & \text{if } \mathrm{cw}_2 < \mathrm{cw}_1 . \end{cases} \tag{17.8.1}$$

Δ'' is finitely generated by (6.17) and obviously $\Delta'' \subset \Delta^{(1+)}$. By (15.8) $\mathrm{cw}_1\,\Delta' = \mathrm{cw}_1\,\Delta^{(1)}$ and thus, $\mathrm{cw}_1\,\Delta'' = \mathrm{cw}_1\,\Delta^{(1+)} = \mathrm{cw}_2$. Finally, Lemma 17.9 below states that $\Delta^{(1+)} \subset \mathrm{Dr}_1\,\Delta''$; this shows that $\Delta^{(1+)}$ is finitely Der-generated. ■

Thus, we have to prove

(17.9) Lemma. $\Delta^{(1+)} \subset \mathrm{Dr}_1\,\Delta''$.

Before we prove this lemma, we need another

(17.10 Lemma. $\Delta^{(1)} \subset u^{\mathrm{mon}_1} \cdot \mathrm{Dr}_1\,\Delta''$.

Proof is given below in (17.13).

(17.11) Proof of Lemma 17.9: (17.10) yields

$$\widetilde{\Delta} = \Delta^{(1)} : u^{\mathrm{mon}_1} \subset \mathrm{Dr}_1\,\Delta'' \tag{17.11.1}$$

which shows that (17.9) is true if $\mathrm{cw}_2 \geq \mathrm{cw}_1$. If $\mathrm{cw}_2 < \mathrm{cw}_1$, then we have to use (17.11.1) together with

$$u^{\frac{\mathrm{cw}_2}{\mathrm{cw}_1 - \mathrm{cw}_2}\cdot \mathrm{mon}_1} \cdot (1) \subset \Delta'' \subset \mathrm{Dr}_1\,\Delta'' \tag{17.11.2}$$

(which is clear from (17.8.1)). It is easy to see that (17.11.1), (17.11.2), and the definition of $\Delta^{(1+)}$ (17.3) show that (17.9) is true if $\mathrm{cw}_2 < \mathrm{cw}_1$. ■

Thus, everything is reduced to Lemma 17.10, and before we prove it, we still need another

(17.12) Lemma. $\{(u^{\mathrm{mon}_1} \cdot \mathrm{Dr}_1\,\Delta'')(\nu/\mathrm{cw}_1)\}$ *is stably contact.*

(17.13) Proof of Lemma 17.10: Clearly, since $\Delta' \subset u^{\mathrm{mon}_1} \cdot (1)$, we have

$$\Delta' = u^{\mathrm{mon}_1} \cdot (\Delta' : u^{\mathrm{mon}_1}) \subset u^{\mathrm{mon}_1} \cdot \Delta'' \subset u^{\mathrm{mon}_1} \cdot \mathrm{Dr}_1\,\Delta''$$

thus,

$$\{\Delta'(\nu/\mathrm{cw}_1)\} \subset \{(u^{\mathrm{mon}_1} \cdot \mathrm{Dr}_1\,\Delta'')(\nu/\mathrm{cw}_1)\}. \tag{17.13.1}$$

At the same time $\{\Delta^{(1)}(\nu/\mathrm{cw}_1)\}$ is the minimal stably contact filtration containing $\{\Delta'(\nu/\mathrm{cw}_1)\}$, and $\{(u^{\mathrm{mon}_1} \cdot \mathrm{Dr}_1\,\Delta'')(\nu/\mathrm{cw}_1)\}$ is stably contact by (17.12) and contains $\{\Delta'(\nu/\mathrm{cw}_1)\}$ by (17.13.1); consequently,

$$\{\Delta^{(1)}(\nu/\mathrm{cw}_1)\} \subset \{(u^{\mathrm{mon}_1} \cdot \mathrm{Dr}_1\,\Delta'')(\nu/\mathrm{cw}_1)\}$$

and this yields (17.10). ■

Now all we have to prove is Lemma 17.12, and this is, of course, the key point of the proof of the finite Der-generatedness of $\Delta^{(1+)}$.

(17.14) Proof of Lemma 17.12: What we have to do is just to apply the Main Lemma 12.3 to the following case. Let

$$\Delta''' = \{(\mathrm{Dr}_1\,\Delta'')(\nu/\mathrm{cw}_2)\}$$

— this is by definition a stably contact filtration in $\mathcal{O}_1$. Let $c = \mathrm{cw}_2/\mathrm{cw}_1$; then

$$\{\Delta'''(c\nu)\} = \left\{(\mathrm{Dr}_1\,\Delta'')\left(\nu \cdot \frac{\mathrm{cw}_2}{\mathrm{cw}_1} \cdot \frac{1}{\mathrm{cw}_2}\right)\right\} = \{(\mathrm{Dr}_1\,\Delta'')(\nu/\mathrm{cw}_1)\}. \tag{17.14.1}$$

Let $\alpha = \mathrm{mon}_1 / \mathrm{cw}_1$; then

$$u^{\alpha} \cdot \{\Delta'''(c\nu)\} = u^{\mathrm{mon}_1 / \mathrm{cw}_1} \cdot \{(\mathrm{Dr}_1 \Delta'')(\nu / \mathrm{cw}_1)\} = \{(u^{\mathrm{mon}_1} \cdot \mathrm{Dr}_1 \Delta'')(\nu / \mathrm{cw}_1)\} . \tag{17.14.2}$$

We want to apply the Main Lemma 12.3 to the filtration Δ''' , the fractional monomial u^{α} , and the number c . In order to do so, we have to check the conditions (12.3.1) and (12.3.2).

(12.3.1) says that $|\alpha| + c > 1$, i.e., in our case $|\mathrm{mon}_1 / \mathrm{cw}_1| + \mathrm{cw}_2 / \mathrm{cw}_1 > 1$, or $|\mathrm{mon}_1| + \mathrm{cw}_2 > \mathrm{cw}_1$; we know this from (17.2).

(12.3.2) says that if $c < 1$ (i.e., $\mathrm{cw}_2 < \mathrm{cw}_1$), then

$$u^{\frac{c}{1-c} \cdot \alpha} \cdot (1) \subset \{\Delta'''(c\nu)\}$$

— this is easy to see from (17.14.1) and (17.8.1).

Thus, the Main Lemma 12.3 yields the stable contactness of the filtration (17.14.2), which is exactly what we need. ■

Chapter 4. Change of the subring

18. The result

(18.1). Let $\mathcal{O}$ be a complete normal crossing ring with the maximal ideal $\mathcal{M}$.

We have shown in (13.15.1) that for any filtration Δ in $\mathcal{O}$ such that $\mathrm{Dr}_1\,\Delta$ is defined,

$$\mathrm{Dr}_1\,\Delta = \left\{(x_1, x_2, \ldots, x_{m_1})^{\mathrm{cw}_1\cdot\nu}\right\} * \Delta^{(1)} \tag{18.1.1}$$

where $\Delta^{(1)}$ is a filtration in a subring $\mathcal{O}_1 \subset \mathcal{O}$, such that $\mathcal{O}$ as a normal crossing ring is represented $\mathcal{O} = \mathcal{O}_1[[x_1, x_2, \ldots, x_{m_1}]]$. We have also shown that the filtration $\Delta^{(1)}$ is uniquely defined by the subring $\mathcal{O}_1$ and the elements $x_1, x_2, \ldots, x_{m_1}$.

In this subsection we study how the filtration $\Delta^{(1)}$ changes as we change the presentation $\mathcal{O} = \mathcal{O}_1[[x_1, x_2, \ldots, x_{m_1}]]$.

(18.2) Remark. Assume we have two different presentations of $\mathcal{O}$ as a formal power series ring

$$\mathcal{O} = \mathcal{O}_1[[x_1, x_2, \ldots, x_{m_1}]] = \mathcal{O}_1'[[x_1', x_2', \ldots, x_{m_1}']] \tag{18.2.1}$$

(both equalities are understood as equalities of normal crossing rings). Assume also that

(18.2.2) The two sets $x_1, x_2, \ldots, x_{m_1}$ and $x_1', x_2', \ldots, x_{m_1}'$ generate the same subspace in $\mathcal{M}/\mathcal{M}^2$.

Then we may consider a map

$$\mathcal{O}_1 \hookrightarrow \mathcal{O} \to \mathcal{O}/(x_1', x_2', \ldots, x_{m_1}') = \mathcal{O}_1' . \tag{18.2.3}$$

It is easy to see that this composion map $\mathcal{O}_1 \xrightarrow{\sim} \mathcal{O}_1'$ is an isomorphism of normal crossing rings.

Similarly, we may consider the other map

$$\mathcal{O}_1' \hookrightarrow \mathcal{O} \to \mathcal{O}/(x_1, x_2, \ldots x_{m_1}) = \mathcal{O}_1 \tag{18.2.4}$$

and it gives an isomorphism $\mathcal{O}_1' \xrightarrow{\sim} \mathcal{O}_1$. Note, however, that these two isomorphisms need not be inverse to each other.

Note also that the fixed variables of $\mathcal{O}_1$ and $\mathcal{O}_1'$ are exactly those fixed variables of $\mathcal{O}$ whose images in $\mathcal{M}/\mathcal{M}^2$ do not lie in the subspace generated by $x_1, x_2, \ldots, x_{m_1}$ (or equivalently, by $x_1', x_2', \ldots, x_m'$). This means that the fixed variables of each of $\mathcal{O}_1$ and $\mathcal{O}_1'$ are in a one-to-one correspondence with the same subset of the fixed variables of $\mathcal{O}$; in particular, the fixed variables of $\mathcal{O}_1$ correspond to the fixed variables of $\mathcal{O}_1'$.

As the isomorphisms $\mathcal{O}_1 \xrightarrow{\sim} \mathcal{O}_1'$ and $\mathcal{O}_1' \xrightarrow{\sim} \mathcal{O}_1$ of (18.2.3) and (18.2.4) preserve the structures of normal crossing rings, they preserve the ordering of the fixed variables; thus, they map each of the fixed variables into the corresponding fixed variable of the other ring (of course, up to multiplication by an invertible element).

(18.3) Proposition. *Suppose we are given a filtration* Δ *in* $\mathcal{O}$ *such that* $\mathrm{Dr}_1 \Delta$ *is defined, and two presentations of* $\mathcal{O}$ *as in (18.2.1). Assume also that*

$$x_1, x_2, \ldots, x_{m_1},\ x'_1, x'_2, \ldots, x'_{m_1} \in (\mathrm{Dr}_1 \Delta)(1/\mathrm{cw}_1)$$

where $\mathrm{cw}_1 = \mathrm{cw}_1 \Delta$ *, and the images of both* $x_1, x_2, \ldots, x_{m_1}$ *and* $x'_1, x'_2, \ldots, x'_{m_1}$ *in* $\mathcal{M}/\mathcal{M}^2$ *form two bases of the subspace*

$$\left[(\mathrm{Dr}_1 \Delta)(1/\mathrm{cw}_1) \bmod \mathcal{M}^2\right] \subset \mathcal{M}/\mathcal{M}^2 .$$

Consider the two presentations (18.1.1)

$$\mathrm{Dr}_1 \Delta = \left\{(x_1, x_2, \ldots, x_{m_1})^{\mathrm{cw}_1 \cdot \nu}\right\} * \Delta^{(1)} = \left\{(x'_1, x'_2, \ldots, x'_{m_1})^{\mathrm{cw}_1 \cdot \nu}\right\} * \Delta_1^{(1)} \tag{18.3.1}$$

where $\Delta^{(1)}$ *and* $\Delta_1^{(1)}$ *are filtrations in* $\mathcal{O}_1$ *and* $\mathcal{O}'_1$ *respectively. Then the isomorphism* $\mathcal{O}_1 \tilde{\to} \mathcal{O}'_1$ *(18.2.3) transforms* $\Delta^{(1)}$ *into* $\Delta_1^{(1)}$ *.*

(18.4) Remark. Interchanging $\mathcal{O}_1$ with $\mathcal{O}'_1$, we find that the isomorphism $\mathcal{O}'_1 \tilde{\to} \mathcal{O}_1$ (18.2.4) transforms $\Delta_1^{(1)}$ into $\Delta^{(1)}$; thus, their composition $\mathcal{O}_1 \tilde{\to} \mathcal{O}'_1 \tilde{\to} \mathcal{O}_1$ may be a nontrivial automorphism of $\mathcal{O}_1$ preserving the filtration $\Delta^{(1)}$.

(18.5) Lemma. *In the notation and the assumptions of (18.2) consider the homomorphism*

$$\begin{aligned} \mathcal{O}_1[[X_1, X_2, \ldots, X_{m_1}]] &\to \mathcal{O} \\ X_i &\to x'_i \end{aligned} \tag{18.5.1}$$

(here $X_1, X_2, \ldots, X_{m_1}$ *are independent variables over* $\mathcal{O}_1$ *). Then this map is an isomorphism, and we can write* $\mathcal{O} = \mathcal{O}_1[[x'_1, x'_2, \ldots, x'_{m_1}]]$ *.*

(18.6) Proof: Indeed, this is a homomorphism of regular local rings of the same dimension, and it is easy to see that the condition (18.2.2) yields that the corresponding map between the tangent spaces at the closed points of the spectra is an isomorphism. This shows that the map (18.5.1) is monomorphic with a dense image; as both rings are complete, this map must be an isomorphism. ■

(18.7) Lemma. *Assume* $\mathcal{O} = \mathcal{O}_1[[x_1, x_2, \ldots, x_m]]$ *as normal crossing rings, and let* Δ, Δ_1 *be two integrally closed filtrations in* $\mathcal{O}, \mathcal{O}_1$ *respectively, such that for some* $c \in \mathbf{Q}$ *,* $c > 0$ *,*

$$\Delta = \left\{(x_1, x_2, \ldots, x_m)^{c \cdot \nu}\right\} * \Delta_1 .$$

Then

$$\Delta_1 = \Delta \cap \mathcal{O}_1 = \left[\Delta \bmod (x_1, x_2, \ldots, x_m)\right] .$$

(18.8) Proof: Indeed, both equalities are obvious from Proposition 5.10. ■

(18.9) Lemma. *In the notation and the assumptions of (18.3)*

$$\mathrm{Dr}_1 \Delta = \left\{(x'_1, x'_2, \ldots, x'_{m_1})^{\mathrm{cw}_1 \cdot \nu}\right\} * \Delta^{(1)} .$$

(18.10) Proof: Indeed, $\mathcal{O} = \mathcal{O}_1[[x'_1, x'_2, \ldots, x'_{m_1}]]$ by (18.5), and

$$x'_1, x'_2, \ldots, x'_{m_1} \in (\mathrm{Dr}_1 \Delta)(1/\mathrm{cw}_1) .$$

Thus, by (13.15.1)

$$\mathrm{Dr}_1\,\Delta = \left\{(x'_1, x'_2, \ldots, x'_{m_1})^{\mathrm{cw}_1\cdot\nu}\right\} * \Delta_2^{(1)}$$

where $\Delta_2^{(1)}$ is a filtration in $\mathcal{O}_1$.

By (18.7), $\Delta_2^{(1)}$ is given by

$$\Delta_2^{(1)} = (\mathrm{Dr}_1\,\Delta) \cap \mathcal{O}_1 \ .$$

At the same time we can apply (18.7) to

$$\mathrm{Dr}_1\,\Delta = \left\{(x_1, x_2, \ldots, x_{m_1})^{\mathrm{cw}_1\cdot\nu}\right\} * \Delta^{(1)} \ .$$

This yields

$$\Delta^{(1)} = \mathrm{Dr}_1\,\Delta \cap \mathcal{O}_1$$

thus, $\Delta_2^{(1)} = \Delta^{(1)}$ and

$$\mathrm{Dr}_1\,\Delta = \left\{(x'_1, x'_2, \ldots, x'_{m_1})^{\mathrm{cw}_1\cdot\nu}\right\} * \Delta^{(1)} \ .$$

■

(18.11) Proof of Proposition 18.3: It is easy to see that the image of the filtration $\Delta^{(1)}$ under the map (18.2.2) is just $\Delta^{(1)} \bmod (x'_1, x'_2, \ldots, x'_{m_1})$, i.e., the filtration in

$$\mathcal{O}'_1 = \mathcal{O}/(x'_1, x'_2, \ldots, x'_m)$$

generated by $\Delta^{(1)}(\nu) \bmod (x'_1, x'_2, \ldots, x'_{m_1})$ for all ν , where $\Delta^{(1)}(\nu) \subset \mathcal{O}_1$ is identified with its image in $\mathcal{O} \supset \mathcal{O}_1$. Thus, we need to prove that

$$[\Delta^{(1)} \bmod (x'_1, x'_2, \ldots, x'_{m_1})] = \Delta_1^{(1)} \ .$$

It is easy to see from Lemma 18.9 that

$$[\Delta^{(1)} \bmod (x'_1, x'_2, \ldots, x'_{m_1})] = [(\mathrm{Dr}_1\,\Delta) \bmod (x'_1, x'_2, \ldots, x'_{m_1})] \tag{18.11.1}$$

(both are considered as filtrations in $\mathcal{O}'_1 = \mathcal{O}/(x'_1, x'_2, \ldots, x'_{m_1})$). Applying (18.7) to

$$\mathrm{Dr}_1\,\Delta = \left\{(x'_1, x'_2, \ldots, x'_{m_1})^{\mathrm{cw}_1\cdot\nu}\right\} * \Delta_1^{(1)}$$

we find

$$[(\mathrm{Dr}_1\,\Delta) \bmod (x'_1, x'_2, \ldots, x'_{m_1})] = \Delta_1^{(1)} \ . \tag{18.11.2}$$

Thus, from (18.11.1) and (18.11.2) we see that

$$[\Delta^{(1)} \bmod (x'_1, x'_2, \ldots, x'_{m_1})] = \Delta_1^{(1)} \tag{18.11.3}$$

and this is what we need. ■

19. The implications

(19.1). We start with the invariance of the monomial mon_1 of (16.6).

Let Δ be a finitely Der-generated filtration in a complete normal crossing ring $\mathcal{O}$ of characteristic zero, and suppose $\Delta \neq (0),(1)$ and $\mathrm{m}_1\,\Delta < \dim\mathcal{O}$.

For any presentation

$$\mathcal{O} = \mathcal{O}_1[[x_1, x_2, \ldots, x_{m_1}]] \tag{19.1.1}$$

$$\mathrm{Dr}_1\,\Delta = \{(x_1, x_2, \ldots, x_{m_1})^{\mathrm{cw}_1 \cdot \nu}\} * \Delta^{(1)} \tag{19.1.2}$$

as in (13.15.1), we constructed in (16.6) the maximal fractional monomial u^{mon_1} dividing $\Delta^{(1)}$.

Note that u^{mon_1} was defined as a fractional monomial in $\mathcal{O}_1$; as the fixed variables of $\mathcal{O}_1$ are clearly u_i, $i \in I_1(\Delta)$, mon_1 is a function on $I_1(\Delta)$. If $\Delta^{(1)} \neq (0)$ then mon_1 is a function

$$\mathrm{mon}_1 : I_1(\Delta) \to \mathbf{Q}_+$$

and if $\Delta^{(1)} = (0)$, then $\mathrm{mon}_1 = \infty$ (i.e., it is a constant function with the value ∞).

Now suppose we change the presentation (19.1.1); then (19.1.2) also changes, so $\Delta^{(1)}$ and (a priori) mon_1 change.

(19.2) Corollary to Proposition 18.3. mon_1 *is independent of the presentation (19.1.1).*

(19.3) Proof: Indeed, as in (18.2.1), take two different presentations of $\mathcal{O}$:

$$\mathcal{O} = \mathcal{O}_1[[x_1, x_2, \ldots, x_{m_1}]] = \mathcal{O}_1'[[x_1', x_2', \ldots, x_{m_1}']] .$$

Then, as in (18.3.1), we get two filtrations $\Delta^{(1)}, \Delta_1^{(1)}$ in $\mathcal{O}_1, \mathcal{O}_1'$ respectively, and Proposition 18.3 states that the isomorphism $\mathcal{O} \tilde{\to} \mathcal{O}_1'$ of (18.2.2) transforms $\Delta^{(1)}$ into $\Delta_1^{(1)}$.

Note that $\mathcal{O}_1$ and $\mathcal{O}_1'$ are normal crossing rings with fixed variables numbered by the same set $I_1(\Delta)$, and the isomorphism (18.2.2) is an isomoprhism of normal crossing rings preserving the numbering of the fixed variables (as we explained in (18.2)).

This, of course, means that the maximal fractional monomial dividing $\Delta^{(1)}$ is mapped into the maximal fractional monomial dividing $\Delta_1^{(1)}$, and the exponents of these monomials (which are functions on $I_1(\Delta)$ as in (19.1)) coincide. This shows that mon_1 is independent of the choice of the presentation (19.1.1), so it is an intrinsic invariant of the filtration Δ. ∎

(19.4) Definition. *We shall write* $\mathrm{mon}_1\,\Delta = \mathrm{mon}_1$; $\mathrm{mon}_1\,\Delta$ *is defined for any finitely* Der-*generated filtration* $\Delta \neq (0),(1)$, *such that* $\mathrm{m}_1\,\Delta < \dim\mathcal{O}$.

(19.5). We retain the notation of (19.1).

Recall that in case $\mathrm{mon}_1 = \mathrm{mon}_1\,\Delta \neq \infty$ we defined in Section 17

$$\mathrm{cw}_2 = \mathrm{cw}_1(\Delta^{(1)} : u^{\mathrm{mon}_1}) \tag{19.5.1}$$

and

$$\Delta^{(1+)} = \begin{cases} \Delta^{(1)} : u^{\mathrm{mon}_1}, & \text{if } \mathrm{cw}_2 \geq \mathrm{cw}_1 \\ (\Delta^{(1)} : u^{\mathrm{mon}_1}) * \left[u^{\frac{\mathrm{cw}_2}{\mathrm{cw}_1 - \mathrm{cw}_2} \cdot \mathrm{mon}_1} \cdot (1) \right], & \text{if } \mathrm{cw}_2 < \mathrm{cw}_1 . \end{cases} \tag{19.5.2}$$

In case $\mathrm{mon}_1 = \mathrm{mon}_1 \Delta = \infty$ (i.e., $\Delta^{(1)} = (0)$) let $\mathrm{cw}_2 = \infty$, $\Delta^{(1+)} = (0)$.

Again, we may ask how these objects depend upon the presentation of $\mathcal{O}$ (19.1.1). Namely, given two presentations of $\mathcal{O}$ as in (18.2.1), we may also define

$$\mathrm{cw}_2' = \mathrm{cw}_1(\Delta_1^{(1)} : u^{\mathrm{mon}_1}) \ . \tag{19.5.3}$$

(Recall that by (19.2) mon_1 is the same for $\Delta^{(1)}$ and $\Delta_1^{(1)}$.) Finally, let

$$\Delta_1^{(1+)} = \begin{cases} \Delta_1^{(1)} : u^{\mathrm{mon}_1} \ , & \text{if } \mathrm{cw}_2' \geq \mathrm{cw}_1 \\ (\Delta_1^{(1)} : u^{\mathrm{mon}_1}) * \left[u^{\frac{\mathrm{cw}_2'}{\mathrm{cw}_1 - \mathrm{cw}_2'} \cdot \mathrm{mon}_1} \cdot (1) \right] \ , & \text{if } \mathrm{cw}_2' < \mathrm{cw}_1 \ . \end{cases} \tag{19.5.4}$$

(Again, if $\mathrm{mon}_1 \Delta = \infty$, let $\mathrm{cw}_2' = \infty$, $\Delta_1^{(1+)} = (0)$.)

(19.6) Corollary to Proposition 18.3. *(i) The number* cw_2 *is independent of the presentation (19.1.1), i.e.,* $\mathrm{cw}_2 = \mathrm{cw}_2'$.

(ii) $\Delta^{(1+)}$ *is independent of the presentation (19.1.1) in the following sense. For any two presentations as in (18.2.1), the isomorphism* $\mathcal{O} \tilde{\to} \mathcal{O}_1'$ *of (18.2.3) maps* $\Delta^{(1+)}$ *into* $\Delta_1^{(1+)}$.

(19.7) Proof: Indeed, part (i) is established by the reasoning completely parallel to (19.3), and after we know $\mathrm{cw}_2 = \mathrm{cw}_2'$, the same reasoning yields (ii). ■

(19.8) Definiton. *We shall write* $\mathrm{cw}_2 \Delta = \mathrm{cw}_2$.

Clearly, $\mathrm{cw}_2 \Delta$ is an intrinsic invariant of the filtration Δ ; it is defined for the same class of filtrations for which $\mathrm{mon}_1 \Delta$ is defined.

(19.9). Now we turn to the results concerning special filtrations. First of all, let Δ be a special filtration given in the form (7.2.3). In (13.24) we have seen that the first two terms in the first characteristic string of Δ are $(\mathrm{cw}_1 \Delta, -\mathrm{m}_1 \Delta)$. Now we see that the next two terms are $|\mathrm{mon}_1| = |\mathrm{mon}_1 \Delta|$ (as clearly $\mathrm{mon}_1 = \mathrm{mon}_1 \Delta$) and $c_2 = \mathrm{cw}_2 \Delta$.

Now let $\Delta \neq (0), (1)$ be a special filtration in the same ring $\mathcal{O}$ as in (19.1), and let $\mathcal{M}$ be the maximal ideal of $\mathcal{O}$. By (13.21) $\mathrm{Dr}_1 \Delta = \Delta$.

Consider a presentation of $\mathcal{O}$ as a normal crossing ring

$$\mathcal{O} = \mathcal{O}_1[[x_1, x_2, \ldots, x_{m_1}]] \tag{19.9.1}$$

where $x_1, x_2, \ldots, x_{m_1}$ are such that their images in $\mathcal{M}/\mathcal{M}^2$ generate the subspace

$$[\Delta(1/\mathrm{cw}_1) \bmod \mathcal{M}^2] = [(\mathrm{Dr}_1 \Delta)(1/\mathrm{cw}_1) \bmod \mathcal{M}^2] \subset \mathcal{M}/\mathcal{M}^2$$

where $\mathrm{cw}_1 = \mathrm{cw}_1 \Delta$. Then by (13.15.1)

$$\Delta = \mathrm{Dr}_1 \Delta = \{(x_1, x_2, \ldots, x_{m_1})^{\mathrm{cw}_1 \cdot \nu}\} * \Delta^{(1)} \tag{19.9.2}$$

where $\Delta^{(1)}$ is a filtration in $\mathcal{O}_1$.

Suppose $m_1 = \mathrm{m}_1 \Delta < \dim \mathcal{O}$. Then, according to (19.2) and (19.4), the maximal monomial dividing $\Delta^{(1)}$ is u^{mon_1} , where $\mathrm{mon}_1 = \mathrm{mon}_1 \Delta$. Let

$$\tilde{\Delta} = \Delta^{(1)} : u^{\mathrm{mon}_1} \tag{19.9.3}$$

(Again, if $\mathrm{mon}_1 = \infty$, i.e., $\Delta^{(1)} = (0)$, then let $\tilde{\Delta} = (0)$.)

(19.10) Corollary to Proposition 18.3. *$\widetilde{\Delta}$ is a special filtration in $\mathcal{O}_1$, and Δ may be obtained from $\widetilde{\Delta}$ according to the rules (7.6) for the inductive construction of the special filtrations. In other words*

$$\Delta = \left\{(x_1, x_2, \ldots, x_{m_1})^{\mathrm{cw}_1 \cdot \nu}\right\} * (u^{\mathrm{mon}_1} \cdot \widetilde{\Delta}) \tag{19.10.1}$$

where

(19.10.2) $|\mathrm{mon}_1| + \mathrm{cw}_1 \widetilde{\Delta} > \mathrm{cw}_1$

(19.10.3) either $\mathrm{cw}_1 \widetilde{\Delta} \geq \mathrm{cw}_1$ *or*

$$u^{\frac{\mathrm{cw}_1 - \mathrm{cw}_1 \widetilde{\Delta}}{\mathrm{cw}_1 \widetilde{\Delta}} \cdot \mathrm{mon}_1} \cdot (1) \subset \widetilde{\Delta}$$

(cf. (7.6.1) and (7.6.2)).

(19.11) Remark. According to (19.6) and (19.8), $\mathrm{cw}_1 \widetilde{\Delta} = \mathrm{cw}_2 \Delta$.

(19.12) Proof of (19.10): As Δ is a special filtration, it can be obtained by the inductive rules of (7.6). This means that Δ can be obtained by suspension of a filtration of the form $u^\alpha \cdot \Delta'$ and rescaling. In other words, we can find some presentation of $\mathcal{O}$ as a normal crossing ring

$$\mathcal{O} = \mathcal{O}_1'[[x_1', x_2', \ldots, x_m']] \tag{19.12.1}$$

and some special filtration $\widetilde{\Delta}'$ in $\mathcal{O}_1'$, such that

$$\Delta = \left\{(x_1', x_2', \ldots, x_m')^{c \cdot \nu}\right\} * (u^\alpha \cdot \widetilde{\Delta}') \tag{19.12.2}$$

for some $c \in \mathbf{Q}$, $c > 0$, and some fractional monomial u^α in $\mathcal{O}_1'$, satisfying the properties

(19.12.3) $|\alpha| + \mathrm{cw}_1 \widetilde{\Delta}' > c$

(19.12.4) either $\mathrm{cw}_1 \widetilde{\Delta}' \geq c$ or

$$u^{\frac{c - \mathrm{cw}_1 \widetilde{\Delta}'}{\mathrm{cw}_1 \widetilde{\Delta}'} \cdot \alpha} \cdot (1) \subset \widetilde{\Delta}'$$

(cf. (7.6.1), (7.6.2)).

It follows immediately from (19.12.2) that

$$\mathrm{cw}_1 \Delta = c \tag{19.12.5}$$

and

$$\mathrm{m}_1 \Delta = m \tag{19.12.6}$$

thus, $c = \mathrm{cw}_1$ and $m = m_1$. Moreover, the images of $x_1', x_2', \ldots, x_{m_1}'$ ($m_1 = m$) clearly generate the subspace

$$[\Delta(1/\mathrm{cw}_1) \bmod \mathcal{M}^2] \subset \mathcal{M}/\mathcal{M}^2 .$$

Now recall that in (19.9) we assumed that $x_1, x_2, \ldots, x_{m_1}$ generate the same subspace. This shows that we can apply Proposition 18.3; thus, the isomorphism $\mathcal{O}_1 \tilde{\to} \mathcal{O}_1'$ (18.2.3) transforms the filtration $\Delta^{(1)} = u^{\mathrm{mon}_1} \cdot \widetilde{\Delta}$ (see (19.9.2), (19.9.3)) into the filtration $u^\alpha \cdot \widetilde{\Delta}'$.

Notice that $\widetilde{\Delta}'$ is special; thus, it is not divisible by any monomial, so u^α is the maximal monomial dividing $u^\alpha \cdot \widetilde{\Delta}'$. Thus, u^{mon_1} is transformed into u^α , i.e., $\alpha = \mathrm{mon}_1$, and $\widetilde{\Delta}$ is transformed into $\widetilde{\Delta}'$; hence, $\widetilde{\Delta}$ is special. Applying the same isomorphism to the properties (19.12.3) and (19.12.4), we get (19.10.2) and (19.10.3). Finally, (19.10.1) follows from (19.9.2) and (19.9.3). ∎

(19.13) Corollary to Proposition 18.3. *Let* Δ *be a special filtration given in the form (7.2.3); then all the numerical data* $n_1, n_2, \ldots, n_k$, $c_1, c_2, \ldots, c_{k+1}$, $\mathrm{mon}_1, \mathrm{mon}_2, \ldots, \mathrm{mon}_k$ *are intrinsically defined by the filtration* Δ . *In other words, if* Δ *is given in the form (7.2.3) in some other coordinate system, say,* $x'_1, x'_2, \ldots, x'_N$, *then the numerical data in both presentations are the same. In particular, the first characteristic string* $\mathrm{char}_1 \Delta$ *is an intrinsic invariant of* Δ .

(19.14) Remark. Clearly, $n_1 = \mathrm{m}_1 \Delta$, $c_1 = \mathrm{cw}_1 \Delta$, $\mathrm{mon}_1 = \mathrm{mon}_1 \Delta$, $c_2 = \mathrm{cw}_2 \Delta$.

(19.15) Proof of (19.13): Indeed, we already know that n_1 , c_1 , and mon_1 are intrinsic invariants of Δ .

Now suppose Δ is given in the form (7.2.3) in two different coordinate systems, say $x_1, x_2, \ldots, x_N$ and $x'_1, x'_2, \ldots, x'_N$. Then clearly

$$\Delta = \{(x_1, x_2, \ldots, x_{n_1})^{c_1 \cdot \nu}\} * (u^{\mathrm{mon}_1} \cdot \widetilde{\Delta})$$

and

$$\Delta = \{(x'_1, x'_2, \ldots, x'_{n_1})^{c_1 \cdot \nu}\} * (u^{\mathrm{mon}_1} \cdot \widetilde{\Delta}')$$

where $\widetilde{\Delta}$ and $\widetilde{\Delta}'$ respectively are special filtrations in

$$\mathcal{O}_1 = \mathbf{k}[[x_{n_1+1}, x_{n_1+2}, \ldots, x_N]]$$

and

$$\mathcal{O}'_1 = \mathbf{k}[[x'_{n_1+1}, x'_{n_1+2}, \ldots, x'_N]] .$$

To show that the numerical invariants of Δ in these two coordinate systems coincide, we need to show that the numerical invariants of $\widetilde{\Delta}$ and $\widetilde{\Delta}'$ coincide. As we have already noted in (19.12), the isomorphism $\mathcal{O}_1 \xrightarrow{\sim} \mathcal{O}'_1$ (18.2.3) maps $\widetilde{\Delta}$ into $\widetilde{\Delta}'$, so induction on N completes the proof. ■

(19.16) Remark. Apparently, there should be some more general result that for any polyhedral filtration its polyhedron is an intrinsic invariant. In particular, this would yield that the numerical invariants of any almost special filtration are intrinsically defined. All our needs, however, are covered by Corollary 19.13.

20. Proof of the Main Theorem 7.11

(20.1) Remark. Recall that for a special filtration Δ given in the form (7.2.3), its first characteristic string is

$$\mathrm{char}_1 \Delta = (c_1, -m_1, |\,\mathrm{mon}_1\,|, c_2, -m_2, |\,\mathrm{mon}_2\,|, \ldots) \tag{20.1.1}$$

(cf. (7.8.1)).

As we noted in (19.9), here $c_1 = \mathrm{cw}_1 \Delta$, $m_1 = \mathrm{m}_1 \Delta$, $\mathrm{mon}_1 = \mathrm{mon}_1 \Delta$, $c_2 = \mathrm{cw}_2 = \mathrm{cw}_2 \Delta$. Thus, the first four terms ($\mathrm{cw}_1 \Delta$, $-\mathrm{m}_1 \Delta$, $|\mathrm{mon}_1 \Delta|$, $\mathrm{cw}_2 \Delta$) may be defined for any finitely Der-generated filtration Δ , provided $\Delta \neq (0), (1)$ and $\mathrm{m}_1 \Delta < \dim \mathcal{O}$. In case $\mathrm{m}_1 \Delta = \dim \mathcal{O}$ only the first two terms ($\mathrm{cw}_1 \Delta$, $-\mathrm{m}_1 \Delta$) are defined, and in case $\Delta = (0)$ or $\Delta = (1)$ only the first term $\mathrm{cw}_1 \Delta$ is defined.

(20.2) Proposition. *If* $\Delta \subset \Delta'$ *are two finitely* Der-*generated filtrations, then*

(20.2.1) $\operatorname{cw}_1 \Delta \geq \operatorname{cw}_1 \Delta'$;

(20.2.2) If $0 < \operatorname{cw}_1 \Delta = \operatorname{cw}_1 \Delta' < \infty$, *then* $-\operatorname{m}_1 \Delta \geq -\operatorname{m}_1 \Delta'$;

(20.2.3) If, in addition, $-\operatorname{m}_1 \Delta = -\operatorname{m}_1 \Delta' > -\dim \mathcal{O}$, *then*

$$|\operatorname{mon}_1 \Delta| \geq |\operatorname{mon}_1 \Delta'| .$$

In particular,

$$(\operatorname{cw}_1 \Delta, -\operatorname{m}_1 \Delta, |\operatorname{mon}_1 \Delta|) \geq (\operatorname{cw}_1 \Delta', -\operatorname{m}_1 \Delta', |\operatorname{mon}_1 \Delta'|) \tag{20.2.4}$$

if all the terms here are defined. (The inequality in (20.2.4) is understood lexicographically, the terms on the left being the most significant.)

(20.3) Proof: Indeed, the statements (20.2.1) and (20.2.2) we already know from (13.2) and (13.17.4).

Now suppose $\operatorname{cw}_1 \Delta = \operatorname{cw}_1 \Delta' = \operatorname{cw}_1$, $0 < \operatorname{cw}_1 < \infty$, and $\operatorname{m}_1 \Delta = \operatorname{m}_1 \Delta' = m_1 < \dim \mathcal{O}$. Let $\mathcal{O} = \mathcal{O}_1[[x_1, x_2, \ldots, x_{m_1}]]$ be a presentation of $\mathcal{O}$ as a normal crossing ring, such that $x_1, x_2, \ldots, x_{m_1} \in (\operatorname{Dr}_1 \Delta)(1/\operatorname{cw}_1)$. By (13.17.4) $\operatorname{Dr}_1 \Delta \subset \operatorname{Dr}_1 \Delta'$, so $x_1, x_2, \ldots, x_{m_1} \in (\operatorname{Dr}_1 \Delta')(1/\operatorname{cw}_1)$, and the images of $x_1, x_2, \ldots, x_{m_1}$ in $\mathcal{M}/\mathcal{M}^2$ form a basis of the subspace

$$\left[(\operatorname{Dr}_1 \Delta)(1/\operatorname{cw}_1) \bmod \mathcal{M}^2\right] = \left[(\operatorname{Dr}_1 \Delta')(1/\operatorname{cw}_1) \bmod \mathcal{M}^2\right] \subset \mathcal{M}/\mathcal{M}^2$$

and thus we have the presentations (13.15.1) for $\operatorname{Dr}_1 \Delta$ and $\operatorname{Dr}_1 \Delta'$:

$$\operatorname{Dr}_1 \Delta = \left\{(x_1, x_2, \ldots, x_{m_1})^{\operatorname{cw}_1 \cdot \nu}\right\} * \Delta^{(1)} \tag{20.3.1}$$

$$\operatorname{Dr}_1 \Delta' = \left\{(x_1, x_2, \ldots, x_{m_1})^{\operatorname{cw}_1 \cdot \nu}\right\} * \Delta_1^{(1)} \tag{20.3.2}$$

where $\Delta^{(1)}$ and $\Delta_1^{(1)}$ are filtrations in $\mathcal{O}_1$. Note that $\Delta^{(1)} \subset \Delta_1^{(1)}$ since $\operatorname{Dr}_1 \Delta \subset \operatorname{Dr}_1 \Delta'$.

Now $u^{\operatorname{mon}_1 \Delta}$ and $u^{\operatorname{mon}_1 \Delta'}$ are the maximal fractional monomials in $\mathcal{O}_1$ dividing $\Delta^{(1)}$ and $\Delta_1^{(1)}$ respectively. As $\Delta^{(1)} \subset \Delta_1^{(1)}$, $u^{\operatorname{mon}_1 \Delta'}$ divides $u^{\operatorname{mon}_1 \Delta}$, hence

$$|\operatorname{mon}_1 \Delta| \geq |\operatorname{mon}_1 \Delta'|$$

which proves (20.2.3) and (20.2.4). ■

(20.4) Proposition. *Suppose there is an equality in (20.2.4). Then*

(20.4.1) $\operatorname{mon}_1 \Delta = \operatorname{mon}_1 \Delta'$

(20.4.2) $\operatorname{cw}_2 \Delta \geq \operatorname{cw}_2 \Delta'$

Let $\operatorname{cw}_1 = \operatorname{cw}_1 \Delta = \operatorname{cw}_1 \Delta'$, $\operatorname{mon}_1 = \operatorname{mon}_1 \Delta = \operatorname{mon}_1 \Delta'$. *If* $\operatorname{mon}_1 = \infty$, *let* $\Delta^{(1+)} = \Delta_1^{(1+)} = (0)$ *(cf. (19.5)). If* $\operatorname{mon}_1 \neq \infty$, *then, as in (17.3), let*

$$\Delta^{(1+)} = \begin{cases} \Delta^{(1)} : u^{\operatorname{mon}_1} , & \text{if } \operatorname{cw}_2 \Delta \geq \operatorname{cw}_1 \\ (\Delta^{(1)} : u^{\operatorname{mon}_1}) * \left[u^{\frac{\operatorname{cw}_2 \Delta}{\operatorname{cw}_1 - \operatorname{cw}_2 \Delta} \cdot \operatorname{mon}_1} \cdot (1)\right] , & \text{if } \operatorname{cw}_2 \Delta < \operatorname{cw}_1 . \end{cases} \tag{20.4.3}$$

$$\Delta_1^{(1+)} = \begin{cases} \Delta_1^{(1)} : u^{\operatorname{mon}_1} , & \text{if } \operatorname{cw}_2 \Delta' \geq \operatorname{cw}_1 \\ (\Delta_1^{(1)} : u^{\operatorname{mon}_1}) * \left[u^{\frac{\operatorname{cw}_2 \Delta'}{\operatorname{cw}_1 - \operatorname{cw}_2 \Delta'} \cdot \operatorname{mon}_1} \cdot (1)\right] , & \text{if } \operatorname{cw}_2 \Delta' < \operatorname{cw}_1 . \end{cases} \tag{20.4.4}$$

Then

(20.4.5) $\Delta^{(1+)} \subset \Delta_1^{(1+)}$.

(20.5) Proof: Indeed, we have already seen in (20.3) that $u^{\mathrm{mon}_1 \Delta'}$ divides $u^{\mathrm{mon}_1 \Delta}$; thus, the equality in (20.2.4), which includes $|\mathrm{mon}_1 \Delta'| = |\mathrm{mon}_1 \Delta|$, yields $\mathrm{mon}_1 \Delta = \mathrm{mon}_1 \Delta'$.

We keep all the notation of (20.3); as we have seen there, $\Delta^{(1)} \subset \Delta_1^{(1)}$. Thus,

$$\Delta^{(1)} : u^{\mathrm{mon}_1} \subset \Delta_1^{(1)} : u^{\mathrm{mon}_1} \tag{20.5.1}$$

and

$$\mathrm{cw}_2 \Delta = \mathrm{cw}_1(\Delta^{(1)} : u^{\mathrm{mon}_1}) \geq \mathrm{cw}_1(\Delta_1^{(1)} : u^{\mathrm{mon}_1}) = \mathrm{cw}_2 \Delta' . \tag{20.5.2}$$

Thus, we have proved (20.4.1) and (20.4.2), and we have to prove (20.4.5).

If $\mathrm{cw}_2 \Delta \geq \mathrm{cw}_1$, then

$$\Delta^{(1+)} = \Delta^{(1)} : u^{\mathrm{mon}_1} \subset \Delta_1^{(1)} : u^{\mathrm{mon}_1} \subset \Delta_1^{(1+)} . \tag{20.5.3}$$

Now assume $\mathrm{cw}_2 \Delta < \mathrm{cw}_1$; then $\mathrm{cw}_2 \Delta' < \mathrm{cw}_1$ as $\mathrm{cw}_2 \Delta' \leq \mathrm{cw}_2 \Delta < \mathrm{cw}_1$. In this case

$$\Delta^{(1+)} =(\Delta^{(1)} : u^{\mathrm{mon}_1}) * \left[u^{\frac{\mathrm{cw}_2 \Delta}{\mathrm{cw}_1 - \mathrm{cw}_2 \Delta} \cdot \mathrm{mon}_1} \cdot (1)\right] \tag{20.5.4}$$

$$\Delta_1^{(1+)} =(\Delta^{(1)} : u^{\mathrm{mon}_1}) * \left[u^{\frac{\mathrm{cw}_2 \Delta'}{\mathrm{cw}_1 - \mathrm{cw}_2 \Delta'} \cdot \mathrm{mon}_1} \cdot (1)\right] . \tag{20.5.5}$$

As $\mathrm{cw}_2 \Delta \geq \mathrm{cw}_2 \Delta'$,

$$\frac{\mathrm{cw}_2 \Delta}{\mathrm{cw}_1 - \mathrm{cw}_2 \Delta} \geq \frac{\mathrm{cw}_2 \Delta'}{\mathrm{cw}_1 - \mathrm{cw}_2 \Delta'} \tag{20.5.6}$$

and

$$u^{\frac{\mathrm{cw}_2 \Delta}{\mathrm{cw}_1 - \mathrm{cw}_2 \Delta} \cdot \mathrm{mon}_1} \cdot (1) \subset u^{\frac{\mathrm{cw}_2 \Delta'}{\mathrm{cw}_1 - \mathrm{cw}_2 \Delta'} \cdot \mathrm{mon}_1} \cdot (1) . \tag{20.5.7}$$

Finally, (20.5.1), (20.5.7), (20.5.4), and (20.5.5) together yield

$$\Delta^{(1+)} \subset \Delta_1^{(1+)}$$

in this case too. ■

(20.6) Remark. We may apply these results (i.e., (20.2) and (20.4)) to the case when Δ' is a special filtration containing Δ ; then we see that the first four terms* of the first characteristic string of Δ' are bounded from above by

$$(\mathrm{cw}_1 \Delta, -\mathrm{m}_1 \Delta, |\mathrm{mon}_1 \Delta|, \mathrm{cw}_2 \Delta) .$$

Moreover, the case when this bound is reached is covered by the following

* By the first four terms here we mean all those of the first four terms that are defined — see (20.1).

(20.7) Proposition. *Let* Δ *be a finitely* Der*-generated filtration in* $\mathcal{O}$ *, such that* $\Delta \neq (0),(1)$ *and* $\mathrm{m}_1\,\Delta < \dim\mathcal{O}$ *, and let* $\mathcal{O}_1, \Delta^{(1)}, \Delta^{(1+)}$ *be as in (20.3), (20.4).*

Consider special filtrations Δ' *in* $\mathcal{O}$ *containing* Δ *and such that*

$$(\mathrm{cw}_1\,\Delta', -\,\mathrm{m}_1\,\Delta', |\,\mathrm{mon}_1\,\Delta'|, \mathrm{cw}_2\,\Delta') = (\mathrm{cw}_1\,\Delta, -\,\mathrm{m}_1\,\Delta, |\,\mathrm{mon}_1\,\Delta|, \mathrm{cw}_2\,\Delta)\ . \tag{20.7.1}$$

Then such filtrations Δ' *are in a one-to-one correspondence with the special filtrations* $\widetilde{\Delta}$ *in* $\mathcal{O}_1$ *containing* $\Delta^{(1+)}$ *and such that* $\mathrm{cw}_1\,\widetilde{\Delta} = \mathrm{cw}_1\,\Delta^{(1+)}$ *. This correspondence is given by*

$$\Delta' = \left\{(x_1, x_2, \ldots, x_{m_1})^{\mathrm{cw}_1 \cdot \nu}\right\} * (u^{\mathrm{mon}_1} \cdot \widetilde{\Delta}) \tag{20.7.2}$$

(20.8) Proof: Indeed, take any special filtration $\Delta' \supset \Delta$, such that it satisfies (20.7.1), and let $\Delta_1^{(1+)}$ be as in (20.4.4). Then by (20.4.5)

$$\Delta_1^{(1+)} \supset \Delta^{(1+)} \tag{20.8.1}$$

and by (17.4)

$$\mathrm{cw}_1\,\Delta_1^{(1+)} = \mathrm{cw}_2\,\Delta' = \mathrm{cw}_2\,\Delta = \mathrm{cw}_1\,\Delta^{(1+)}\ . \tag{20.8.2}$$

On the other hand, (19.10.3) shows that in our case

$$\Delta_1^{(1+)} = \Delta_1^{(1)} : u^{\mathrm{mon}_1} \tag{20.8.3}$$

i.e.,

$$\Delta' = \left\{(x_1, x_2, \ldots, x_{m_1})^{\mathrm{cw}_1 \cdot \nu}\right\} * (u^{\mathrm{mon}_1} \cdot \Delta_1^{(1+)}) \tag{20.8.4}$$

and also (19.10) yields that $\Delta_1^{(1+)}$ is special. Thus, we may take $\widetilde{\Delta} = \Delta_1^{(1+)}$ to be the special filtration containing $\Delta^{(1+)}$, which corresponds to Δ' .

Conversely, take any special filtration $\widetilde{\Delta}$ in $\mathcal{O}$, such that $\widetilde{\Delta} \supset \Delta^{(1+)}$ and $\mathrm{cw}_1\,\widetilde{\Delta} = \mathrm{cw}_1\,\Delta^{(1+)}$; as we have already seen in (20.8.2), $\mathrm{cw}_1\,\Delta^{(1+)} = \mathrm{cw}_2\,\Delta$, so

$$\mathrm{cw}_1\,\widetilde{\Delta} = \mathrm{cw}_2\,\Delta\ . \tag{20.8.5}$$

Now take Δ' given by (20.7.2); then Δ' is an almost special filtration in $\mathcal{O}$. To show that Δ' is special, we need to show that the conditions (7.6.1) and (7.6.2) are satisfied. In our case these conditions amount to the following:

(20.8.6) $|\,\mathrm{mon}_1\,| + \mathrm{cw}_1\,\widetilde{\Delta} > \mathrm{cw}_1$;

(20.8.7) either $\mathrm{cw}_1\,\widetilde{\Delta} \geq \mathrm{cw}_1$ or

$$u^{\frac{\mathrm{cw}_1\,\widetilde{\Delta}}{\mathrm{cw}_1 - \mathrm{cw}_1\,\widetilde{\Delta}} \cdot \mathrm{mon}_1} \cdot (1) \subset \widetilde{\Delta}\ .$$

Applying (17.2) to Δ , we get

$$|\,\mathrm{mon}_1\,| + \mathrm{cw}_2\,\Delta \geq \mathrm{cw}_1\ . \tag{20.8.8}$$

As $\mathrm{cw}_2\,\Delta = \mathrm{cw}_1\,\widetilde{\Delta}$ by (20.8.5), this yields (20.8.6).

According to the way $\Delta^{(1+)}$ was defined in (20.4.3), the following statement is true:

(20.8.9) either $\mathrm{cw}_2\,\Delta \geq \mathrm{cw}_1$ or

$$u^{\frac{\mathrm{cw}_2\,\Delta}{\mathrm{cw}_1 - \mathrm{cw}_2\,\Delta}\cdot \mathrm{mon}_1} \cdot (1) \subset \Delta^{(1+)} .$$

As $\Delta^{(1+)} \subset \widetilde{\Delta}$ and $\mathrm{cw}_2\,\Delta = \mathrm{cw}_1\,\widetilde{\Delta}$, this yields (20.8.7). Thus, Δ' is a special filtration, and it is clear from the construction that it satisfies the condition (20.7.1). Thus, Δ' can be taken as a special filtration containing Δ, which corresponds to $\widetilde{\Delta}$.

Finally, it is easy to see that these two constructions ($\widetilde{\Delta}$ from Δ' and Δ' from $\widetilde{\Delta}$) are inverse to each other, so the correspondence between them is one-to-one. ■

(20.9) Remark. Consider two special filtrations Δ' and $\widetilde{\Delta}$ in the rings $\mathcal{O}$ and $\mathcal{O}_1$ respectively, which correspond to each other according to (20.7); then their first characteristic strings are related in the following way :

(20.9.1) $$\mathrm{char}_1\,\Delta' = (\mathrm{cw}_1\,\Delta, -\,\mathrm{m}_1\,\Delta, |\,\mathrm{mon}_1\,\Delta|, \mathrm{char}_1\,\widetilde{\Delta})$$

In other words, to get $\mathrm{char}_1\,\Delta'$ from $\mathrm{char}_1\,\widetilde{\Delta}$, we should insert three extra terms $(\mathrm{cw}_1\,\Delta, -\,\mathrm{m}_1\,\Delta, |\,\mathrm{mon}_1\,\Delta|)$ in the beginning.

Now we are ready for

(20.10) Proof of the Main Theorem 7.11: We apply induction on $\dim \mathcal{O}$ to prove the following statement:

(20.10.1) *Given a finitely* Der-*generated integrally closed filtration* Δ *in a complete normal crossing ring* $\mathcal{O}$ *of characteristic zero, there exists a unique special filtration* Δ^* *in* $\mathcal{O}$, *such that it contains* Δ *and* $\mathrm{char}_1\,\Delta^*$ *is the greatest among the first characteristic strings of all special filtrations containing* Δ. *In addition,* $\mathrm{cw}_1\,\Delta^* = \mathrm{cw}_1\,\Delta$, *and if* Δ *is already special, then* $\Delta^* = \Delta$.

This statement is stronger than the Main Theorem 7.11 in two aspects: first, it allows Δ to be finitely Der-generated (and not only finitely generated), and second, it asserts that $\mathrm{cw}_1\,\Delta^* = \mathrm{cw}_1\,\Delta$.

The base of our inductive argument is the trivial case $\dim \mathcal{O} = 0$ — then $\mathcal{O}$ is a field and all integrally closed filtrations in $\mathcal{O}$ (i.e., just (0) and (1)) are special.

Now we turn to the inductive step. Take any finitely Der-generated filtration Δ in $\mathcal{O}$ ($\mathcal{O}$ being as in (20.10.1)), and assume the statement (20.10.1) is proven for all filtrations in the rings of dimension less than $\dim \mathcal{O}$; we need to prove the statement (20.10.1) for the filtration Δ.

First consider the cases $\Delta = (0)$ and $\Delta = (1)$. If $\Delta = (1)$, then $\Delta^* = (1)$ is the only special filtration containing Δ. If $\Delta = (0)$, then $\Delta^* = (0)$ is the minimal special filtration containing Δ, and its first characteristic string (which has only one term $\mathrm{cw}_1\,\Delta = \mathrm{cw}_1\,\Delta^* = \infty$) is the greatest among the first characteristic strings of all special filtrations. Thus, in both of these cases Δ^* exists and is unique, and $\Delta^* = \Delta$. In particular, $\mathrm{cw}_1\,\Delta^* = \mathrm{cw}_1\,\Delta$.

Now assume $\Delta \neq (0), (1)$, i.e., $0 < \mathrm{cw}_1\,\Delta < \infty$, and denote $\mathrm{cw}_1 = \mathrm{cw}_1\,\Delta$, $m_1 = \mathrm{m}_1\,\Delta$. If $m_1 = \dim \mathcal{O}$, then

$$\mathrm{Dr}_1\,\Delta = \{\mathcal{M}^{\mathrm{cw}_1 \cdot \nu}\}$$

and $\Delta^* = \mathrm{Dr}_1\,\Delta = \{\mathcal{M}^{\mathrm{cw}_1 \cdot \nu}\}$ is the special filtration containing Δ, whose first characteristic string is the greatest. Indeed, if Δ' is any other special filtration, $\Delta' \supset \Delta$, then $\mathrm{cw}_1\,\Delta' \leq \mathrm{cw}_1\,\Delta = \mathrm{cw}_1$, and if $\mathrm{cw}_1\,\Delta' = \mathrm{cw}_1$, then (13.21) and (13.25) yield

$$\{\mathcal{M}^{\mathrm{cw}_1 \cdot \nu}\} \supset \Delta' = \mathrm{Dr}_1\,\Delta' \supset \mathrm{Dr}_1\,\Delta = \{\mathcal{M}^{\mathrm{cw}_1 \cdot \nu}\}$$

i.e., $\Delta' = \{\mathcal{M}^{\mathrm{cw}_1 \cdot \nu}\}$. Note that in this case also $\mathrm{cw}_1 \Delta = \mathrm{cw}_1 \Delta^*$, and if Δ is special, then $\Delta = \Delta^*$.

Now assume $0 < \mathrm{cw}_1 < \infty$ and $m_1 < \dim \mathcal{O}$. Take

$$\mathcal{O} = \mathcal{O}_1[[x_1, x_2, \ldots, x_{m_1}]] \tag{20.10.2}$$

$$\Delta = \{(x_1, x_2, \ldots, x_{m_1})^{\mathrm{cw}_1 \cdot \nu}\} * \Delta^{(1)} \tag{20.10.3}$$

as in (13.15.1), and take $\Delta^{(1+)}$ as in (20.4.3). By Proposition 17.7 $\Delta^{(1+)}$ is a finitely Der-generated filtration in $\mathcal{O}_1$. As by (14.4.3) $\dim \mathcal{O}_1 < \dim \mathcal{O}$, we may apply our inductive assumption, and it yields the existence of a unique special filtration $\Delta^{(1+)*}$ in $\mathcal{O}_1$, $\Delta^{(1+)*} \supset \Delta^{(1+)}$, whose first charactersitic string is maximal among the first characteristic strings of the special filtrations containing $\Delta^{(1+)}$. In addition, the inductive assumption yields $\mathrm{cw}_1 \Delta^{(1+)*} = \mathrm{cw}_1 \Delta^{(1+)}$.

By (20.7) there is a correspondence between special filtrations in $\mathcal{O}$ containing Δ , and special filtrations in $\mathcal{O}_1$ containing $\Delta^{(1+)}$; the special filtrations in $\mathcal{O}$ should satisfy (20.7.1), and the special filtrations $\widetilde{\Delta}$ in $\mathcal{O}_1$ should satisfy $\mathrm{cw}_1 \widetilde{\Delta} = \mathrm{cw}_1 \Delta^{(1+)}$. As the special filtration $\Delta^{(1+)*}$ satisfies $\mathrm{cw}_1 \Delta^{(1+)*} = \mathrm{cw}_1 \Delta^{(1+)}$, it is included in this correspondence; let Δ^* be the corresponding special filtration in $\mathcal{O}$; thus,

$$\Delta^* = \{(x_1, x_2, \ldots, x_{m_1})^{\mathrm{cw}_1 \cdot \nu}\} * (u^{\mathrm{mon}_1} \cdot \Delta^{(1+)*}) \tag{20.10.4}$$

$$\Delta^* \supset \Delta \tag{20.10.5}$$

$$\begin{aligned}(\mathrm{cw}_1 \Delta^*, -\mathrm{m}_1 \Delta^*, |\mathrm{mon}_1 \Delta^*|, \mathrm{cw}_1 \Delta^*) &= (\mathrm{cw}_1 \Delta, -\mathrm{m}_1 \Delta, |\mathrm{mon}_1 \Delta|, \mathrm{cw}_2 \Delta) \\ &= (\mathrm{cw}_1, -m_1, |\mathrm{mon}_1|, \mathrm{cw}_1 \Delta^{(1+)}) \ .\end{aligned} \tag{20.10.6}$$

We claim Δ^* is the special filtration containing Δ with the maximal first characteristic string. Indeed, we know that Δ^* is special and $\Delta^* \supset \Delta$. Let Δ' be any other special filtration containing Δ . By (20.6) the first four terms of $\mathrm{char}_1 \Delta'$ are bounded from above by

$$(\mathrm{cw}_1 \Delta, -\mathrm{m}_1 \Delta, |\mathrm{mon}_1 \Delta|, \mathrm{cw}_2 \Delta)$$

which by (20.10.6) coincides with the first four terms of $\mathrm{char}_1 \Delta^*$; thus, either

$$\mathrm{char}_1 \Delta' < \mathrm{char}_1 \Delta^* \tag{20.10.7}$$

or the first four terms of $\mathrm{char}_1 \Delta'$ and $\mathrm{char}_1 \Delta^*$ coincide, i.e.,

$$(\mathrm{cw}_1 \Delta', -\mathrm{m}_1 \Delta', |\mathrm{mon}_1 \Delta'|, \mathrm{cw}_2 \Delta') = (\mathrm{cw}_1 \Delta, -\mathrm{m}_1 \Delta, |\mathrm{mon}_1 \Delta|, \mathrm{cw}_2 \Delta) \ . \tag{20.10.8}$$

In the latter case condition (20.7.1) is satisfied, so Δ' corresponds to some special filtration $\widetilde{\Delta}$ in $\mathcal{O}_1$, $\widetilde{\Delta} \supset \Delta^{(1+)}$, and $\widetilde{\Delta}$ is related to Δ' by (20.7.2).

Now $\Delta^{(1+)*}$ is the special filtration containing $\Delta^{(1+)}$ with the maximal first characteristic string, and $\widetilde{\Delta}$ is another special filtration containing $\Delta^{(1+)}$; thus,

$$\mathrm{char}_1 \widetilde{\Delta} \leq \mathrm{char}_1 \Delta^{(1+)*} \tag{20.10.9}$$

and the equality here is possible only if $\widetilde{\Delta} = \Delta^{(1+)*}$.

Δ^* and Δ' are special filtrations in $\mathcal{O}$ which correspond to the special filtrations $\Delta^{(1+)*}$ and $\widetilde{\Delta}$ in $\mathcal{O}_1$; thus, by (20.9)

$$\operatorname{char}_1 \Delta = (\operatorname{cw}_1 \Delta, -\operatorname{m}_1 \Delta, |\operatorname{mon}_1 \Delta|, \operatorname{char}_1 \Delta^{(1+)*}) \tag{20.10.10}$$

$$\operatorname{char}_1 \Delta' = (\operatorname{cw}_1 \Delta, -\operatorname{m}_1 \Delta, |\operatorname{mon}_1 \Delta|, \operatorname{char}_1 \widetilde{\Delta}) . \tag{20.10.11}$$

Thus, from (20.10.9)–(20.10.11) we see that

$$\operatorname{char}_1 \Delta' \leq \operatorname{char}_1 \Delta^* \tag{20.10.12}$$

and the equality here is possible only if $\widetilde{\Delta} = \Delta^{(1+)*}$, i.e., $\Delta' = \Delta^*$. This shows that Δ^* is the unique special filtration containing Δ with the maximal first characterstic string; in addition, $\operatorname{cw}_1 \Delta = \operatorname{cw}_1 \Delta^*$ by (20.10.6).

Finally, if Δ is special, then $\Delta^{(1+)}$ is also special; hence, $\Delta^{(1+)*} = \Delta^{(1+)}$ by induction, so $\Delta = \Delta^*$.

■

Conclusion

21. Some unsolved problems

Here we would like to propose some conjectures about the possibility of coordinate-free definition of Newton polyhedra of more general form than our special polyhedra (i.e., the polyhedra corresponding to our special filtrations).

Let $\mathcal{O}$ be a complete regular local ring. If $x_1, x_2, \ldots, x_N$ is a coordinate system in $\mathcal{O}$ ($N = \dim \mathcal{O}$), and $\Gamma \subset \mathbf{Q}_+^N$ — a polyhedron, then we defined in Section 1 the ideal x^Γ , which is generated by all the monomials

$$x_1^{\alpha_1} x_2^{\alpha_2} \ldots x_N^{\alpha_N}$$

where $(\alpha_1, \alpha_2, \ldots, \alpha_N) \in \mathbf{Z}_+^N \cap \Gamma$.

(21.1) Definition. *We say that the ideal x^Γ is a **polyhedral ideal**; we also say that it is polyhedral in the coordinate system*

$$x = (x_1, x_2, \ldots, x_N) .$$

We should note here that a polyhedral ideal may be polyhedral in different coordinate systems.

(21.2) Conjecture. *Suppose $\dim \mathcal{O} = 2$, and consider all polyhedral ideals containing a given element $f \in \mathcal{O}$. Then there are finitely many polyhedral ideals $\mathcal{A}_1, \mathcal{A}_2, \ldots, \mathcal{A}_s$ containing f , such that any other polyhedral ideal x^Γ containing f , contains one of these ideals $\mathcal{A}_i$, $1 \le i \le s$, and $\mathcal{A}_i$ and x^Γ are polyhedral in one coordinate system.*

(21.3) Motivation for Conjecture 21.2. Let the initial form of f have degree d , and suppose that for some coordinate system (x, y) the expansion of f in the powers of x, y does not include the term x^d . It is not hard to see that in this case the curve $f = 0$ has an analytic branch in the direction of the x-axis. Moreover, if the coefficients in x^k in the expansion of f vanish for all $k \le N$, where N is some number greater than d , then the x-axis is in some sense tangent to a branch of the curve $f = 0$ to some higher degree determined by the Newton polygon of f .

This shows that the Newton polygon of f becomes smaller, if we arrange the coordinate axes tangent to a high degree to the branches of the curve $f = 0$. For each branch, there is a natural limit for this degree of tangency determined by its Puiseaux expansion (for a smooth branch, this limit is infinity — we can arrange the axis along the smooth branch).

Now we see that in order to minimize the Newton polygon, we have to arrange the coordinate axes in such a way that each one is tangent to one of the branches to the highest degree possible for that branch.

Of course, there are only finitely many such arrangements up to the higher order terms, and that is why we could expect that there are only finitely many minimal polyhedral ideals containing f .

Now consider the case $\dim \mathcal{O} > 2$. Then the coordinate systems which minimize the Newton polyhedron of a function f have the property that their coordinate planes (of all dimensions, from hyperplanes to coordinate axes) are tangent to the maximal possible degree to the hypersurface $f = 0$ (or its components or singular strata).

For example, consider the condition of tangency (to the maximal possible degree) of a coordinate axis to a cone. This, of course, means that the axis should lie in the cone. However, it is easy to see that if all smooth lines lying in a cone, form a family, then this family is not even a finite-parameter one ("there are functional parameters"). This shows that we cannot even expect the minimal polyhedral ideals containing a given function f , to form a finite-parameter family.

However, if we try to minimize the polyhedron and not the ideals, it would be natural to expect that the minimal polyhedra would be only finitely many.

(21.4) Conjecture. *Suppose* $\dim \mathcal{O} > 2$ *and* $f \in \mathcal{O}$. *Consider all polyhedral ideals in* $\mathcal{O}$ *containing* f *, and consider all those which are minimal among them (i.e., are not contained in any other of them). Then the corresponding polyhedra are only finitely many. In other words, there are finitely many polyhedra* $\Gamma_1, \Gamma_2, \ldots, \Gamma_s$ *such that the Newton polyhedron of* f *with respect to any coordinate system contains one of* $\Gamma_1, \Gamma_2, \ldots, \Gamma_s$ *, and in addition, each of* $\Gamma_1, \Gamma_2, \ldots, \Gamma_s$ *is the Newton polyhedron of* f *with respect to some coordinate system (although these coordinate systems may be different for each of* $\Gamma_1, \Gamma_2, \ldots, \Gamma_s$ *).*

(21.5) Remark. Both conjectures may be easily generalized for the case of a finite number of functions $f_1, f_2, \ldots, f_n \in \mathcal{O}$ with the weights $\nu_1, \nu_2, \ldots, \nu_n$ — they generate an integrally closed filtration Δ in $\mathcal{O}$, and we may consider all polyhedral filtrations containing Δ . Then we may conjecture that if $\dim \mathcal{O} = 2$, then there are finitely many minimal polyheral filtrations containing Δ , and if $\dim \mathcal{O} > 2$, then the polyhedra corresponding to the minimal polyhedral filtrations containing Δ , are only finitely many.

(21.6) Example. Now we shall show that the Main Theorem 7.11 becomes false if we drop the mysterious property (7.5.2) from the definition of special filtration.

Let $\mathcal{O} = \mathbf{k}[[x, y, u]]$ and suppose $\mathcal{O}$ has only one fixed variable u . Let $n = 1$, $f = x^2 + yu^2$, $\nu_1 = 1$. Then one can see that the special filtration Δ^* with the maximal first characteristic string and such that $f \in \Delta^*(1)$, is given by

$$\Delta^* = \{x^{2\nu}\} * [u^2 \cdot \{y^\nu\}] * [u^4 \cdot \{1\}]$$

In this case $c_1 = 2$, $u^{\mathrm{mon}_1} = u^2$, $c_2 = 1$, $u^{\mathrm{mon}_2} = u^2$, $c_3 = 0$.

Now let us see what happens if we drop the condition (7.5.2) from the definition of a special filtration. It is easy to see that the following filtration

$$\Delta_1 = \{x^{2\nu}\} * [u^2 \cdot \{y^\nu\}]$$

satisfies all the other properties of special filtrations (besides (7.5.2)) and $f \in \Delta_1(1)$. Clearly, the first characteristic string of Δ_1 is minimal among all filtrations with these properties.

On the other hand, for each $\lambda \in \mathbf{k}$ we can make a coordinate change

$$\widetilde{x} = x + \lambda u^2$$

$$\widetilde{y} = y - 2\lambda x - \lambda^2 u^2$$

— then $\mathbf{k}[[x,y,u] = \mathbf{k}[[\widetilde{x},\widetilde{y},u]]$ and $f = x^2 + yu^2 = \widetilde{x}^2 + \widetilde{y}u^2$. Hence, the following filtration

$$\Delta_2 = \{\widetilde{x}^{2\nu}\} * [u^2 \cdot \{\widetilde{y}^\nu\}]$$

has the same properties as Δ_1 . However, it is easy to see that $\Delta_1 \neq \Delta_2$; indeed,

$$\Delta_1(1) = (x^2, xyu, yu^2)$$

$$\Delta_2(1) = (\widetilde{x}^2, \widetilde{x}\widetilde{y}u, \widetilde{y}u^2)$$

and $\Delta_1(1) \neq \Delta_2(1)$.

This shows that we indeed lose the uniqueness in the Main Theorem 7.11 if we drop the condition (7.5.2) from the definition of special filtration.

What we see here, is that the nonuniqueness is due to some quasihomogeneous coordinate changes which appear in continuous families; moreover, a combinatorial condition on the polyhedron eliminates the possibility for such changes. It would be a plausible suggestion that the same coordinate changes account for the possibility that a minimal polyhedral ideal containing a function f , can be varied without changing the corresponding polyhedron.

(21.7) Conjecture. There is a combinatorial condition on the polyhedron ("a rigidity condition") which generalizes (7.5.2); a polyhedron satisfying this condition may appear in only finitely many minimal polyhedral ideals containing a given function f .

Appendices

Logically each appendix belongs in its respective place in the main body of the paper, so that the references to the results of the appendix may be after, but not before that place. Thus, A1 belongs right after Proposition 2.7, A2 after Section 2, A3 after Proposition 5.10, A4 after Proposition 6.9, and A5 after Proposition 11.5.

A1. Weights of a quashihomogeneous filtration are independent of the coordinate system

Here we prove Proposition 2.7. Our way is to show that the weights of a quasihomogeneous filtration can be defined intrinsically, i.e., without reference to any coordinate system. We do not require the rings to be of characteristic zero or even equicharacteristic.

(A1.1) Lemma. *Let $\mathcal{O}$ be a regular local ring with the maximal ideal $\mathcal{M}$, $\mathcal{O}/\mathcal{M} = \mathbf{k}$, and let Δ be a quasihomogeneous filtration in $\mathcal{O}$ with the weights $(w_1, w_2, \ldots, w_n)$. Suppose that all the weights $w_1, w_2, \ldots, w_n$ are positive; then the $\mathcal{O}$-module*

$$\frac{\Delta(\nu)}{\bigcup\limits_{\nu'>\nu} \Delta(\nu')} \tag{A1.1.1}$$

is annihilated by $\mathcal{M}$ and its dimension as a vector space over $\mathbf{k}$ is given by

$$\dim \frac{\Delta(\nu)}{\bigcup\limits_{\nu'>\nu} \Delta(\nu')} = \#\Big\{ (\alpha_1, \alpha_2, \ldots, \alpha_n) \in \mathbf{Z}_+^n \;\Big|\; \sum \alpha_i w_i = \nu \Big\} \tag{A1.1.2}$$

where $\#$ means the cardinality of a set.

(A1.2) Proof: Let $x_1, x_2, \ldots, x_n \in \mathcal{M}$, $\mathcal{M} = (x_1, x_2, \ldots, x_n)$, be a coordinate system which makes Δ quasihomogeneous with the weights $(w_1, w_2, \ldots, w_n)$. Then $x_i \in \Delta(1/w_i)$ and $\mathcal{M} \subset \Delta(c)$ where

$$c = \min_{1 \le i \le n} \frac{1}{w_i}.$$

Thus,

$$\mathcal{M}\Delta(\nu) \subset \Delta(c)\Delta(\nu) \subset \Delta(\nu + c) \subset \bigcup_{\nu'>\nu} \Delta(\nu')$$

which means that the $\mathcal{O}$-module (A1.1.1) is annihilated by $\mathcal{M}$.

$\Delta(\nu)$ is by definition generated by the monomials in $x_1, x_2, \ldots, x_n$ of the weight $\geq \nu$; however, all the monomials of the weight $> \nu$ lie in

$$\bigcup_{\nu'>\nu} \Delta(\nu'),$$

so the module (A1.1.1) is generated by all the monomials of the weight ν exactly. Clearly, these monomials are linearly independent over $\mathbf{k}$, so the dimension of (A1.1.1) is equal to the number of these monomials, i.e., to the right-hand side of (A1.1.2). ■

(A1.3) Proof of (2.7): First consider the case when all the ideals $\Delta(\nu)$ are $\mathcal{M}$-primary; this means that all the weights of Δ are positive in any coordinate system. In this case denote

$$\ell(\nu) = \dim \frac{\Delta(\nu)}{\bigcup_{\nu'>\nu} \Delta(\nu)} . \tag{A1.3.1}$$

Then $\ell(\nu)$ is defined intrinsically by Δ and at the same time by (A1.1.2) $\ell(\nu)$ contains much information about $(w_1, w_2, \dots, w_n)$. Indeed, assume $w_1 \geq w_2 \geq \dots \geq w_n$; then clearly

$$\frac{1}{w_1} = \min\{\nu \in \mathbf{Q}_+ \mid \nu > 0 , \quad \ell(\nu) > 0\} \tag{A1.3.2}$$

and for any i , $2 \leq i \leq n$

$$\frac{1}{w_i} = \min\left\{\nu \in \mathbf{Q}_+ \,\middle|\, \nu > 0 , \quad \ell(\nu) > \#\left\{(\alpha_1, \alpha_2, \dots, \alpha_{i-1}) \in \mathbf{Z}_+^{i-1} \,\middle|\, \sum_{j=1}^{i-1} \alpha_j w_j = \nu\right\}\right\} . \tag{A1.3.3}$$

Thus, we can find w_1 from (A1.3.2) and then $w_2, w_3, w_4, \dots, w_n$ — inductively from (A1.3.3).

Now consider the general case when the ideals $\Delta(\nu)$ are not necessarily $\mathcal{M}$-primary and some of the weights may be equal to zero. Let

$$\wp = \bigcup_{\nu>0} \Delta(\nu) .$$

If Δ is quasihomogeneous with respect to the coordinate system $\{x_1, x_2, \dots, x_n\}$ with the weights $w_1, w_2, \dots, w_n$, where $w_1, w_2, \dots, w_k$ are positive and the rest are zero, then

$$\wp = (x_1, x_2, \dots, x_k) .$$

In particular, this means that the number of zero weights is equal to the depth of the ring $\mathcal{O}/\wp$ and it is an intrinsic invariant of Δ . As for the other weights, localising at $\wp$, we eliminate the zero weights and reduce the question to the first case. ■

A2. Criterion for integral closedness

Here we prove

(A2.1) Proposition. *Let Δ be a decreasing filtration in a ring $\mathcal{O}$, numbered by nonnegative rationals. Assume Δ satisfies the following conditions:*

(A2.1.1) $\Delta(0) = \mathcal{O}$;

(A2.1.2) $\Delta(\nu_1) \cdot \Delta(\nu_2) \subset \Delta(\nu_1 + \nu_2)$;

(A2.1.3) For any $\nu \in \mathbf{Q}_+$

$$\bigcap_{\substack{\nu' \in \mathbf{Q}_+ \\ \nu' < \nu}} \Delta(\nu') = \Delta(\nu)$$

("continuity");

(A2.1.4) If $f^n \in \Delta(n\nu)$, $\nu \in \mathbf{Q}_+$, *then* $f \in \Delta(\nu)$.

Then Δ is integrally closed.

(A2.2) Remark. This criterion shows that in such cases the difference between solving the equations of general form (2.1.3) and the special form (A2.1.4) is not so much as one could expect; this difference is covered by the continuity condition (A2.1.3). I do not know whether the continuity condition (A2.1.3) can be dropped in Proposition A2.1. Note also that the assumptions of Proposition A2.1 mean that the subring $R_t(\Delta) \subset \mathcal{O}[t^{\mathbf{Q}_+}]$ (see Section 3) is completely integrally closed in $\mathcal{O}[t^{\mathbf{Q}_+}]$. (For the notion of completely integrally closed ring, see [Bourbaki], Chapter V, §1, 4°.)

An example of a filtration which does not satisfy the continuity condition (A2.1.3), is given by $\{\mathcal{M}^{\nu+0}\}$.

(A2.3) Proof of Proposition A2.1: We have to prove only that Δ satisfies property (2.1.3). Let $f, g_1, g_2, \ldots, g_n \in \mathcal{O}$, $g_i \in \Delta(i \cdot \nu)$, and assume

(A2.3.1) $$f^n + g_1 f^{n-1} + g_2 f^{n-2} + \ldots + g_n = 0$$

Suppose $f \in \Delta(\nu')$, $0 \le \nu' < \nu$. Then

$$g_1 f^{n-1} \in \Delta(\nu + (n-1)\nu') = \Delta(n\nu' + (\nu - \nu'))$$
$$g_2 f^{n-2} \in \Delta(2\nu + (n-2)\nu') = \Delta(n\nu' + 2(\nu - \nu'))$$

$$g_{n-1} f \in \Delta((n-1)\nu + \nu') = \Delta(n\nu' + (n-1)(\nu - \nu'))$$
$$g_n \in \Delta(n\nu) = \Delta(n\nu' + n(\nu - \nu'))$$

Thus,

$$f^n = -g_1 f^{n-1} - g_2 f^{n-2} - \ldots - g_n \in \Delta(n\nu' + (\nu - \nu'))$$

The property (A2.1.4) yields

$$f \in \Delta\left(\frac{n\nu' + (\nu - \nu')}{n}\right) = \Delta\left(\nu' + \frac{\nu - \nu'}{n}\right)$$

Thus, if $f \in \Delta(\nu')$, $0 \le \nu' < \nu$, then

$$f \in \Delta\left(\nu' + \frac{\nu - \nu'}{n}\right).$$

This means that in fact $f \in \Delta(\nu')$ for any $\nu' < \nu$, i.e.,

$$f \in \bigcap_{\nu' < \nu} \Delta(\nu').$$

By (A2.1.3) $f \in \Delta(\nu)$. ■

A3. The structure of suspension

(A3.1). Here we prove Proposition 5.10 which asserts that the ν-th term of the suspension $\{(x_1,\ldots,x_m)^\nu\} * \Delta$ is equal to

$$\sum_{k=0}^{[\nu]+1} (x_1, x_2, \ldots, x_m)^k \ \Delta(\nu - k) \ .$$

Clearly, it is enough to prove the case $m = 1$ only — the general case will then follow by induction. Thus, we have to show that the 1-fold suspension of an integrally closed filtration Δ in a ring $\mathcal{O}$ coincides with the filtration

$$\Delta'(\nu) = \sum_{k=0}^{[\nu]+1} x^k \cdot \Delta(\nu - k) \tag{A3.1.1}$$

in $\mathcal{O}[[x]]$. By (5.9) the suspension of Δ contains Δ', so it is enough to prove that Δ' is integrally closed, i.e., it satisfies the properties (2.1.1)–(2.1.3). However, Δ' obviously satisfies (2.1.1) and (2.1.2), so we have to prove only that Δ' satisfies the integral closedness condition (2.1.3), i.e., that if $f, g_1, g_2, \ldots, g_n \in \mathcal{O}[[x]]$, $g_i \in \Delta'(i\nu)$, and

$$f^n + g_1 f^{n-1} + g_2 f^{n-2} + \ldots + g_n = 0 \tag{A3.1.2}$$

then $f \in \Delta'(\nu)$.

(A3.2) Lemma. *If the filtration Δ'' in the polynomial ring $\mathcal{O}[x]$, $\Delta''(\nu) = \Delta'(\nu) \cap \mathcal{O}[x]$, is integrally closed (where Δ' is given by (A3.1.1)), then Δ' is integrally closed too.*

(A3.3) Proof: Take $f, g_1, g_2, \ldots, g_n \in \mathcal{O}[[x]]$, $g_i \in \Delta'(i\nu)$, satisfying (A3.1.2); we need to show that $f \in \Delta'(\nu)$.

Let

$$f = \sum_{j=0}^{\infty} f_j x^j$$

where $f_j \in \mathcal{O}$. Clearly, $f \in \Delta'(\nu)$ is equivalent to $f_j \in \Delta(\nu - j)$ for $0 \le j \le \nu$ only. Thus, let

$$\tilde{f} = \sum_{j=0}^{[\nu]} f_j x^j \in \mathcal{O}[x] \ .$$

Then $f - \tilde{f} \in (x^{[\nu]+1}) \subset \Delta'(\nu)$, so $\tilde{f}$ also satisfies an equation

$$\tilde{f}^n + \tilde{g}_1 \tilde{f}^{n-1} + \tilde{g}_2 \tilde{f}^{n-2} + \ldots + \tilde{g}_n = 0$$

where

$$\tilde{g}_i \in \Delta'(i\nu) \ , \quad \tilde{g}_i = \sum_{j=0}^{\infty} \tilde{g}_{ij} x^j \ , \quad \tilde{g}_{ij} \in \mathcal{O} \ .$$

Finally, let

$$\tilde{\tilde{g}}_i = \sum_{j=0}^{[n\nu]} \tilde{g}_{ij} x^j = \mathcal{O}[x]$$

for $i = 1, 2, \ldots, n-1$, and

$$\tilde{\tilde{g}}_n = -\tilde{f}^{n-1} - \tilde{\tilde{g}}_1 \tilde{f}^{n-2} - \ldots - \tilde{\tilde{g}}_{n-1} \tilde{f} \in \mathcal{O}[x] \ .$$

Then clearly $\tilde{\tilde{g}}_i \in \Delta''(i\nu)$ for $i = 1, 2, \ldots, n-1$, and $\tilde{\tilde{g}}_n \in \Delta''(n\nu)$ since $\tilde{\tilde{g}}_n \equiv \tilde{g}_n \pmod{x^{[n\nu]+1}}$. Finally, by our choice of $\tilde{\tilde{g}}_n$

$$\tilde{f}^n + \tilde{\tilde{g}}_1 \tilde{f}^{n-1} + \tilde{\tilde{g}}_2 \tilde{f}^{n-2} + \ldots + \tilde{\tilde{g}}_n = 0$$

and integral closedness of Δ'' yields $\tilde{f} \in \Delta''(\nu)$, and consequently $f \in \Delta'(\nu)$. Thus, Δ' is integrally closed too. ■

To finish the proof of Proposition 5.10, we need only

(A3.4) Lemma. *If Δ is integrally closed in $\mathcal{O}$, then Δ'' is integrally closed in $\mathcal{O}[x]$, where*

$$\Delta''(\nu) = \sum_{k=0}^{\infty} x^k \Delta(\nu - k) \subset \mathcal{O}[x].$$

(A3.4) Proof: Let

$$R_t(\Delta'') = \sum_{\nu \in \mathbf{Q}_+} \Delta''(\nu) t^\nu + \mathcal{O}[x, t^{\mathbf{Q}_-}] = \sum_{\nu \in \mathbf{Q}} \Delta''(\nu) t^\nu \subset \mathcal{O}[x, t^{\mathbf{Q}}].$$

Let $f \in \mathcal{O}[x]$ satisfy an equation (A3.1.2) with $g_i \in \Delta''(i\nu)$, $i = 1, 2, \ldots, n$. Then $ft^\nu \in \mathcal{O}[x, t^{\mathbf{Q}}]$ is integral over $R_t(\Delta'')$.

However,

$$R_t(\Delta'') = \sum_{\nu \in \mathbf{Q}} \sum_{k \in \mathbf{Z}_+} \Delta(\nu - k) x^k t^\nu$$

and this shows that $R_t(\Delta'')$ is $\mathbf{Z}_+$-graded by the powers of x. The ambient ring $\mathcal{O}[x, t^{\mathbf{Q}_+}]$ is also graded in the same way, so by [Bourbaki], Ch. V, §1, Proposition 20 each homogeneous component $f_k x^k t^\nu$ of ft^ν, is integral over $R_t(\Delta'')$. However, it is easy to see that if $f_k x^k t^\nu$, $\nu \geq k$, is integral over $R_t(\Delta'')$, then $f_k t^{\nu-k}$ is integral over $R_t(\Delta)$, and by integral closedness of Δ, $f_k \in \Delta(\nu - k)$. (If $\nu < k$, then $f_k \in \Delta(\nu - k)$ trivially.) Thus, we see that $f = \sum f_k x^k$, $f_k \in \Delta(\nu - k)$, and this is what we needed. ■

A4. Multiplication and division by monomials: Proof of the structure theorem

(A4.1). Here we prove Proposition 6.9 which asserts that

$$(u^\alpha \cdot \Delta)(\nu) = \{f \in \mathcal{O} \mid f^n \in u^{n\nu\alpha} \cdot \Delta(n\nu), \quad n = \operatorname{denom}(\nu\alpha)\} \tag{A4.1.1}$$

and

$$(\Delta : u^\alpha)(\nu) = \{f \in \mathcal{O} \mid u^{n\nu\alpha} \cdot f^n \in \Delta(n\nu), \quad n = \operatorname{denom}(\nu\alpha)\} \tag{A4.1.2}$$

where $\mathcal{O}$ is a normal crossing ring, u^α a fractional monomial in it, and Δ an integrally closed filtration in $\mathcal{O}$.

As we have already noted in (6.10), the right-hand sides of (A4.1.1) and (A4.1.2) are contained in the left-hand sides. Moreover, it is not hard to see from the definitions that the left-hand sides are the minimal integrally closed filtrations containing the right-hand sides. It is also easy to see that the right-hand sides are decreasing filtrations of $\mathcal{O}$ by subsets. Thus, it is enough to prove the following:

(A4.1.3) The right-hand sides of (A4.1.1) and (A4.1.2) are additive subgroups;

(A4.1.4) These filtrations by subgroups satisfy the properties (2.1.1) and (2.1.2) (in particular, this means that these filtrations by subgroups are indeed filtrations by ideals);

(A4.1.5) They satisfy the integral closedness property (2.1.3).

(A4.2) Proof that the right-hand sides are additive subgroups: Let f_1, f_2 lie in the right-hand side of (A4.1.1) for some $\nu \in \mathbf{Q}_+$. Let $n = \operatorname{denom}(\nu\alpha)$. Then $f_1^n, f_2^n \in u^{n\nu\alpha}\Delta(n\nu)$, and it is enough for us to show that $(f_1 + f_2)^n \in u^{n\nu\alpha}\Delta(n\nu)$.

Indeed,

$$(f_1 + f_2)^n = \sum_{k=0}^{n} \frac{n!}{k!(n-k)!} f_1^k f_2^{n-k}$$

so it is enough to show that $f_1^k f_2^{n-k} \in u^{n\nu\alpha}\Delta(n\nu)$.

Let $f_i^n = u^{n\nu\alpha} g_i$, $i = 1, 2$, so that $g_1, g_2 \in \Delta(n\nu)$. Then

$$(f_1^k f_2^{n-k})^n = u^{n^2\nu\alpha} g_1^k g_2^{n-k} ,$$

and

$$\left(\frac{f_1^k f_2^{n-k}}{u^{n\nu\alpha}}\right)^n = g_1^k g_2^{n-k} \in \Delta(n^2\nu) .$$

Since $\mathcal{O}$ is regular, it is integrally closed, and

$$\frac{f_1^k f_2^{n-k}}{u^{n\nu\alpha}} \in \mathcal{O} .$$

Then the integral closedness of Δ yields

$$\frac{f_1^k f_2^{n-k}}{u^{n\nu\alpha}} \in \Delta(n\nu) ,$$

thus,

$$f_1^k f_2^{n-k} \in u^{n\nu\alpha}\Delta(n\nu)$$

and this is what we need. The case of $\Delta : u^\alpha$ is completely parallel, so it is left to the reader. ∎

(A4.3) Proof that the right-hand sides satisfy the multiplicativity properties (2.1.1) and (2.1.2): Indeed, (2.1.1) is obvious. To see (2.1.2), take f_1, f_2 that lie in the right-hand side of (A4.1.1) for $\nu = \nu_1$ and $\nu = \nu_2$ respectively. Let $n_i = \operatorname{denom}(\nu_i\alpha)$, $i = 1, 2$. Then $f_i^{n_i} \in u^{n_i\nu\alpha} \cdot \Delta(n_i\nu)$, so

$$(f_1 f_2)^{n_1 n_2} \in u^{n_1 n_2(\nu_1+\nu_2)\alpha}\, \Delta(n_1 n_2(\nu_1 + \nu_2)) .$$

Let $n = \operatorname{denom}[(\nu_1 + \nu_2)\alpha]$; clearly, n divides $n_1 n_2$, so $n_1 n_2 = nn'$, where n' is a positive integer. Then

$$\left[\frac{(f_1 f_2)^n}{u^{n(\nu_1+\nu_2)\alpha}}\right]^{n'} = \frac{(f_1 f_2)^{n_1 n_2}}{u^{n_1 n_2(\nu_1+\nu_2)\alpha}} \in \Delta(n_1 n_2(\nu_1 + \nu_2)) = \Delta(n' \cdot n(\nu_1 + \nu_2)) .$$

Again by integral closedness of $\mathcal{O}$ we get

$$\frac{(f_1 f_2)^n}{u^{n(\nu_1+\nu_2)\alpha}} \in \mathcal{O}$$

and integral closedness of Δ yields

$$\frac{(f_1 f_2)^n}{u^{n(\nu_1+\nu_2)\alpha}} \in \Delta(n(\nu_1 + \nu_2)) ,$$

so $(f_1 f_2)^n \in u^{n(\nu_1+\nu_2)\alpha}\Delta(n(\nu_1 + \nu_2))$. Thus, the right-hand side of (A4.1.1) satisfies (2.1.2). The case of $\Delta : u^\alpha$ (A4.1.2) is completely similar, so it is left to the reader. ∎

(A4.4) The integral closedness property (2.1.3): This time we shall check it for the right-hand side of (A4.1.2). Denote the filtration in the right-hand side of (A4.1.2) by Δ'. Let $f \in \mathcal{O}$ satisfy an equation

$$f^m + g_1 f^{m-1} + g_2 f^{m-2} + \ldots + g_m = 0 \tag{A4.4.1}$$

where $g_i \in \Delta'(i\nu)$. Let $n = \text{denom}(\nu\alpha)$, and let

$$R_t(\Delta') = \sum_{\nu \in \mathbf{Q}} \Delta'(\nu) t^\nu \subset \mathcal{O}[t^{\mathbf{Q}}] .$$

Then (A4.4.1) means that $ft^\nu \in \mathcal{O}[t^{\mathbf{Q}}]$ is integral over $R_t(\Delta')$; then $(ft^\nu)^n = f^n t^{n\nu}$ is also integral over $R_t(\Delta')$. Then, as one can easily see, f^n satisfies an equation

$$(f^n)^{\widetilde{m}} + \widetilde{g}_1 (f^n)^{\widetilde{m}-1} + \widetilde{g}_2 (f^n)^{\widetilde{m}-2} + \ldots + \widetilde{g}_{\widetilde{m}} = 0$$

where $\widetilde{g}_i \in \Delta'(in\nu)$. However, $\text{denom}(in\nu\alpha) = 1$, so by the definition of Δ' (it is the right-hand side of (A4.1.2)) $u^{in\nu\alpha} \cdot \widetilde{g}_i \in \Delta(in\nu)$. Denote $\widetilde{\widetilde{g}}_i = u^{in\nu\alpha} \cdot \widetilde{g}_i$, then $u^{n\nu\alpha} \cdot f^n$ satisfies the equation

$$(u^{n\nu\alpha} \cdot f^n)^{\widetilde{m}} + \widetilde{\widetilde{g}}_i \cdot (u^{n\nu\alpha} \cdot f^n)^{\widetilde{m}-1} + \widetilde{\widetilde{g}}_2 \cdot (u^{n\nu\alpha} \cdot f^n)^{\widetilde{m}-2} + \ldots + \widetilde{\widetilde{g}}_{\widetilde{m}} = 0$$

So by integral closedness of Δ we get $u^{n\nu\alpha} \cdot f^n \in \Delta(n\nu)$ and thus $f \in \Delta'(\nu)$. This indeed proves that Δ' (which is the right-hand side of (A4.1.2)) satisfies the integral closedness condition (2.1.3). The case of $u^\alpha \cdot \Delta$ (A4.1.1) is again completely parallel and it is left to the reader. ■

A5. Suspension of a stably contact filtration is stably contact: proof of (11.5)

(A5.1) Lemma. *Let Δ be an integrally closed filtration in any ring $\mathcal{O}$; we shall also consider it as a filtration in $\mathcal{O}[[x_1, x_2, \ldots, x_n]]$. Let Q be a quasihomogeneous filtration in $\mathcal{O}[[x_1, x_2, \ldots, x_m]]$ with the weights $w_1, w_2, \ldots, w_n$ (i.e., it is the integrally closed filtration generated by $x_1, x_2, \ldots, x_n$ with the weights $w_1, w_2, \ldots, w_n$). Then*

$$(Q * \Delta)(\nu) = \sum_{\alpha = (\alpha_1, \alpha_2, \ldots, \alpha_n) \in \mathbf{Z}_+^n} x_1^{\alpha_1} x_2^{\alpha_2} \ldots x_n^{\alpha_n} \Delta\left(\nu - \sum w_i \alpha_i\right) . \tag{A5.1.1}$$

(A5.2) Remark. Clearly, in case $w_1 = w_2 = \ldots = w_n = 1$

$$Q = \left\{(x_1, x_2, \ldots, x_n)^\nu\right\}$$

and $Q * \Delta$ is just the suspension of Δ. In this case the statement of Lemma A5.1 is exactly the statement of Proposition 5.10, so (A5.1) is a generalization of (5.10).

(A5.3) Proof of (A5.1): Clearly,

$$Q = \left\{(x_1)^{\nu/w_1}\right\} * \left\{(x_2)^{\nu/w_2}\right\} * \ldots * \left\{(x - n)^{\nu/w_n}\right\}$$

so

$$Q * \Delta = \left\{(x_1)^{\nu/w_1}\right\} * \left\{(x_2)^{\nu/w_2}\right\} * \left\{(x_n)^{\nu/w_n}\right\} * \Delta$$

and induction on n reduces the statement of (A5.1) to the case $n=1$. However, in case $n=1$ we may apply the rescaling, and it reduces the statement to the case $w_1 = 1$, $n=1$. Finally, by (A5.2) this case follows from (5.10). ■

(A5.4) Remark. From now on, as in (11.5), we shall assume $\mathcal{O}$ and $\mathcal{O}_1$ to be complete normal crossing rings of characteristic zero, such that $\mathcal{O} = \mathcal{O}_1[[x_1, x_2, \dots, x_n]]$ as normal crossing rings, and let $\widehat{\mathcal{O}}_1 = \mathcal{O}_1[\sqrt{u}]$, $\widehat{\mathcal{O}} = \mathcal{O}[\sqrt{u}]$. Note that we can consider $\widehat{\mathcal{O}}$ as a formal power series ring over $\widehat{\mathcal{O}}_1$, if we allow fractional powers of those of $x_1, x_2, \dots, x_n$ that are fixed variables, with the following two conditions:

(A5.4.1) In each power series the exponents of all the monomials have a common denominator;

(A5.4.2) In each power series all the coefficients (which a priori are elements of $\widehat{\mathcal{O}}_1$) lie in a finite extension of $\mathcal{O}_1$.

(A5.5) Lemma. *Let Δ' be an integrally closed filtration in $\widehat{\mathcal{O}}_1$ and let Δ'' be a filtration in $\widehat{\mathcal{O}}$ given by*

$$\Delta''(\nu) = \sum_{\alpha=(\alpha_1,\alpha_2,\dots,\alpha_n)} x_1^{\alpha_1} x_2^{\alpha_2} \dots x_n^{\alpha_n}\, \Delta'(\nu - \textstyle\sum \alpha_i) \tag{A5.5.1}$$

where the summation is taken over such $\alpha = (\alpha_1, \alpha_2, \dots, \alpha_n) \in \mathbf{Q}_+^n$ that $x_i^{\alpha_i}$ makes sense in $\widehat{\mathcal{O}}$ for each i, and infinite sums are allowed within the limits of (A5.4.1) and (A5.4.2). Then Δ'' is integrally closed.

(A5.6) Proof: Clearly, it is enough to take a system of finite extensions of $\mathcal{O}$ whose union is $\widehat{\mathcal{O}}$, and show that for each of these extensions $\mathcal{O}'$ the filtration $\Delta'' \cap \mathcal{O}'$ is integrally closed in $\mathcal{O}'$. We can take

$$\mathcal{O}' = \mathcal{O}_1'[[x_1^{1/k_1}, x_2^{1/k_2}, \dots, x_n^{1/k_n}]]$$

where $\mathcal{O}_1' \subset \widehat{\mathcal{O}}_1$ is a finite extension of $\mathcal{O}_1$ and k_i are positive integers, $k_i = 1$ if x_i is not a fixed variable; then $\widehat{\mathcal{O}}$ is the union of all such $\mathcal{O}'$.

Thus, we have to show that $\Delta'' \cap \mathcal{O}'$ is an integrally closed filtration in $\mathcal{O}'$. Let $\Delta_1' = \Delta' \cap \mathcal{O}_1'$; then Δ_1' is an integrally closed filtration in $\mathcal{O}_1'$ and

$$(\Delta'' \cap \mathcal{O}')(\nu) = \sum_{\alpha=(\alpha_1,\alpha_2,\dots,\alpha_n)\in \mathbf{Z}_+} x_1^{\alpha_1/k_1} x_2^{\alpha_2/k_2} \dots x_n^{\alpha_n/k_n} \cdot \Delta_1'(\nu - \textstyle\sum \frac{\alpha_i}{k_i}) . \tag{A5.6.1}$$

Lemma A5.1 shows that the right-hand side of (A5.6.1) is exactly $Q * \Delta_1'$ where Q is the quasihomogeneous filtration in $\mathcal{O}[[x_1^{1/k_1}, x_2^{1/k_2}, \dots, x_n^{1/k_n}]]$ with the weights $1/k_1, 1/k_2, \dots, 1/k_n$. Thus, it is integrally closed (recall that we defined the join $Q * \Delta$ as the minimal integrally closed filtration containing both Q and Δ). Thus, $\Delta'' \cap \mathcal{O}'$ is integrally closed for each $\mathcal{O}'$, and so is Δ''. ■

From now on let Δ_1 be a contact filtration in $\mathcal{O}_1$, and let $\Delta = \{(x_1, x_2, \dots, x_n)^\nu\} * \Delta_1$ be its suspension in $\mathcal{O}$. Let $\widehat{\Delta}$, $\widehat{\Delta}_1$ respectively be the filtrations in $\widehat{\mathcal{O}}$, $\widehat{\mathcal{O}}_1$ which are minimal, integrally closed, satisfy (11.2.1), and contain Δ, Δ_1 respectively.

(A5.7) Lemma.

$$\widehat{\Delta}(\nu) = \sum_{\alpha_1,\alpha_2,\dots,\alpha_n} x_1^{\alpha_1} x_2^{\alpha_2} \dots x_n^{\alpha_n} \widehat{\Delta}_1(\nu - \textstyle\sum \alpha_i) \tag{A5.7.1}$$

where, as in (A5.5), we allow α_i to be rational provided that $x_i^{\alpha_i}$ makes sense in $\widehat{\mathcal{O}}$, and we allow infinite sums within the limits of (A5.4.1) and (A5.4.2).

(A5.8) Proof: Consider the filtration whose ν-th term is given by the right-hand side of (A5.7.1). By (A5.5) it is an integrally closed filtration. As $\widehat{\Delta}_1$ satisfies the condition (11.2.1), the right-hand side of (A5.7.1) also satisfies this condition. It is also clear that the right-hand side of (A5.7.1) contains $\Delta = \{(x_1, x_2, \ldots, x_n)^\nu\} * \Delta_1$. Finally, the right-hand side of (A5.7.1) is clearly the minimal filtration in $\widehat{\mathcal{O}}$ having all these properties — thus, it should coincide with $\widehat{\Delta}$. ∎

(A5.9) Proof of Proposition 11.5: Indeed, let now Δ_1 be stably contact, and let Δ, $\widehat{\Delta}_1$, $\widehat{\Delta}$ be as above.

Thus, Δ is a suspension of Δ_1, and we have to show that Δ is also stably contact, i.e., $\widehat{\Delta} \cap \mathcal{O} = \Delta$.

Using (A5.7), we get

$$\widehat{\Delta} \cap \mathcal{O} = \{(x_1, x_2, \ldots, x_n)^\nu\} * (\widehat{\Delta}_1 \cap \mathcal{O}) .$$

Since $\widehat{\Delta}_1$ is stably contact, $\widehat{\Delta}_1 \cap \mathcal{O} = \Delta_1$ and

$$\widehat{\Delta} \cap \mathcal{O} = \{(x_1, x_2, \ldots, x_n)^\nu\} * \Delta_1 = \Delta .$$

∎

References

[Abhyankar] S. S. Abhyankar, *Weighted expansions for canonical desingularization*, Lecture notes in mathematics, **910**, Springer, Berlin and New York, 1982.

[Bierstone, Milman] E. Bierstone, P. Milman, *Uniformization of analytic spaces*, Journal of the Amer. Math. Soc. (to appear).

[Bourbaki] N. Bourbaki, *Commutative Algebra*, Hermann, Paris, 1972.

[Giraud 1] J. Giraud, *Sur la theorie du contact maximal,* Math. Zeit. **137** (1974), pp. 285–310.

[Giraud 2] J. Giraud, *Etude locale des singularités,* Publ. Math. d'Orsay 1971/72, réimpression 1980.

[Giraud 3] J. Giraud, *Contact maximal en caractéristique positive,* Ann. Scient. Ec. Norm. Sup., 4e Série, **8** (1975), pp. 201–234.

[Hironaka 1] H. Hironaka, *Resolution of singularities of an algeblraic variety over a field of characteristic zero,* Ann. of Math. **79** (1964), pp. 109–326.

[Hironaka 2] H. Hironaka, *Idealistic exponents of singularity,* Algebraic geometry, The Johns Hopkins Centennial Lectures, Johns Hopkins University Press, 1977, pp. 52–125.

[Hironaka 3] H.Hironaka, *Characterisic polyhedra of singularities,* J.Math. Kyoto Univ. **7** (1967), pp. 251–293.

[Hironaka 4] H. Hironaka, *Desingularization of excellent surfaces* (Notes by Bruce Bennett), Lecture Notes in Mathematics **1101**, pp. 99–132, Springer, 1984.

[Milman] P. D. Milman, unpublished lecture notes, 1978.

[Moh] T. T. Moh, *On a stability theorem for local uniformization in characteristic* p , Publ. RIMS, Kyoto Univ. **23** (1987) pp. 965–973.

[Narasimhan] R. Narasimhan, *Hyperplanarity of the singular locus,* Proceedings of the Amer. Math. Soc. **87** (1983), pp. 403–408.

[Youssin 1] B. Youssin, *Newton polyhedra of ideals* (in this volume).

[Youssin 2] B.Youssin, *Canonical formal uniformization in characteristic zero,* in preparation.

Boris Youssin
Harvard University
Department of Mathematics
Cambridge, MA 02138

Current address (1989/90) :
Institute of Mathematics
Hebrew University
Givat-Ram
Jerusalem, Israel

NEWTON POLYHEDRA OF IDEALS

Boris Youssin*

1. Introduction.

In our paper [Youssin 1] which appears in this volume, we introduced a concept of polyhedral filtration in a regular local ring. Given functions $f_1, f_2, \ldots, f_n \in \mathcal{O}$ (here $\mathcal{O}$ is a complete regular local ring of characteristic zero) and positive rational weights $\nu_1, \nu_2, \ldots, \nu_n$, we constructed (see [Youssin 1], (7.11)) a canonical polyhedral filtration Δ such that $f_i \in \Delta(\nu_i)$; this filtration Δ was characterized by a certain minimality property (see details in [Youssin 1], (7.11); see also (2.9) below). We understand this construction as a coordinate-free definition of the Newton polyhedron of this system of functions.

In particular, given a hypersurface germ, this construction yields a polyhedral filtration in the ambient local ring; this is the case when we are given only one function ($n = 1$).

In this paper we extend this result to the nonhypersurface case. Namely, given an ideal $\mathcal{A} \subset \mathcal{O}$, we construct a canonical polyhedral filtration Δ^* in $\mathcal{O}$ which is characterized by the same minimality property (see below in (2.9)); Δ^* also has the property that there exists some base $\{f_1, f_2, \ldots, f_n\}$ of the ideal $\mathcal{A}$ such that $f_i \in \Delta^*(\nu_i)$, where $\nu_i = \nu(f_i)$ is the multiplicity of f_i (i.e. $f_i \in \mathcal{M}^{\nu_i} \setminus \mathcal{M}^{\nu_i+1}$, where $\mathcal{M}$ is the maximal ideal of $\mathcal{O}$).

More precisely, we first construct a canonical *contact* filtration (see the definition in [Youssin 1], (9.2)) which we denote Δ_0, such that there exists a standard base (see [Hironaka 1], p. 208, or Section 2 below) $\{f_1, f_2, \ldots, f_n\}$ of $\mathcal{A}$ such that $f_i \in \Delta_0(\nu_i)$; Δ_0 also has some minimality property (see (2.7)). After that we use the results of [Youssin 1] to construct the canonical polyhedral filtration Δ^* described above; the exact formulation of our results is in (2.7) and (2.9).

The main part of the job is the construction of the filtration Δ_0, and it goes as follows: we first differentiate the ideal $\mathcal{A}$ in some delicate way and get the ideals that we call *generalized Fitting ideals*, then take the minimal contact filtration containing these ideals with appropriate weights — this is done in (7.10). Finally we prove that this filtration has all the properties of the filtration Δ_0. In the Appendix, we present another proof: we first construct the filtration Δ_0 as the minimal contact filtration such that $f_i \in \Delta(\nu_i)$ for some *completely* normalized standard base $f_1, f_2, \ldots, f_n$. Then we show that (i) this definition is independent of the choice of the base, and (ii) the filtration has the required properties.

We believe that this result would provide a good algebraic framework for the theory of maximal contact (see [Hironaka 2], [Giraud 1]); this, however, will not be developed in this paper (although see (7.12) below).

Received by the editors November 28, 1988, and, in revised form, September 13, 1989

* Supported by NSF grant DMS-8806731 at Courant Institute of Mathematical Sciences, New York University.

The application of our result that we have in mind, is the canonical resolution of singularities in characteristic zero; in a subsequent paper [Youssin 2] we shall use the polyhedral filtration Δ^* constructed here, to give an explicit construction of the blowing-up center (i.e. the resolution algorithm) and the measure of singularity that would strictly improve under the blowing-ups.

This paper is a continuation of [Youssin 1], so the knowledge of [Youssin 1] is assumed everywhere; in addition, in the Appendix we use the generalized Weierstrass preparation theorem of [Hironaka 4], Section 4. Besides that, the paper is self-contained, requiring no previous knowledge of any works on resolution on singularities. This is done for the reader's convenience only; many of the ideas and techniques we are using here, are due to previous works (which, I hope, are quoted as appropriate). In particular, the idea of applying differential calculus in the resolution of singularities is due to [Giraud 1]; our *generalized Fitting ideals* (Section 4) are generalizations of a similar notion (with the same name) of [Villamayor 1, 2]. Our concept of *generic position* (Section 6) is the *transversality* of [Giraud 1], and the normalized standard bases we use here are a slight generalization of those introduced by [Hironaka 1]. The *completely normalized standard bases* of the Appendix are due to [Bierstone, Milman 1].

As in [Youssin 1], we consider only the case of complete rings, leaving the incomplete rings for a subsequent paper.

It is my pleasant duty to acknowledge here the impact of my discussions with Edward Bierstone, Heisuke Hironaka, Pierre Milman and Mark Spivakovsky.

2. Standard bases and the main result

2.1 General assumptions: All the rings in this paper will be assumed commutative, with identity and Noetherian.

We start with a treatment of standard bases that follows [Hironaka 1].

2.2 Notation. Let $\mathcal{O}$ be a local ring with a maximal ideal $\mathcal{M}$, and let

$$\operatorname{gr}_n \mathcal{O} = \mathcal{M}^n/\mathcal{M}^{n+1}$$

$$\operatorname{gr} \mathcal{O} = \bigoplus_{n=0}^{\infty} \operatorname{gr}_n \mathcal{O}$$

Let $\mathcal{A} \subset \mathcal{O}$ be an ideal; we denote by $\operatorname{Init} \mathcal{A}$ the ideal of the initial forms of the elements of $\mathcal{A}$. Then $\operatorname{Init} \mathcal{A} \subset \operatorname{gr} \mathcal{O}$ and $\operatorname{Init} \mathcal{A}$ is a homogeneous ideal in $\operatorname{gr} \mathcal{O}$, i.e.,

$$\operatorname{Init} \mathcal{A} = \bigoplus_{n=0}^{\infty} \operatorname{Init}_n \mathcal{A}.$$

If $f \in \mathcal{M}^n \setminus \mathcal{M}^{n+1}$, we shall write $n = \nu(f)$.

2.3 Proposition ([Hironaka 1], Corollary on p. 208). *Let $f_1, f_2, \ldots, f_n$ be such elements of $\mathcal{A}$ that their initial forms $\overline{f}_1, \overline{f}_2, \ldots \overline{f}_n \in \operatorname{gr} \mathcal{O}$ form a base of $\operatorname{Init} \mathcal{A}$. Then $f_1, f_2, \ldots, f_n$ form a base of $\mathcal{A}$.*

Proof follows easily from Artin-Rees lemma and Nakayama's lemma, so it is left to the reader. ■

2.4 Remark. The homogeneous ideal $\operatorname{Init} \mathcal{A}$ in the graded ring $\operatorname{gr} \mathcal{O}$ has a minimal base. Indeed, for each degree n the linear space $\operatorname{Init}_n \mathcal{A}$ has a subspace of elements lying in the ideal generated by $\operatorname{Init}_k \mathcal{A}$ for all $k < n$, and we can take a basis of any complementary subspace. Taking such bases for all n, we clearly get a base of $\operatorname{Init} \mathcal{A}$, which we shall call a *minimal* base.

It is easy to see that this class of bases is indeed minimal as any homogeneous base of Init $\mathcal{A}$ contains some minimal base. It is also easy to see that any two minimal bases of Init $\mathcal{A}$ have the same number of elements of each degree. In other words, the degrees $\nu_1, \nu_2, \ldots, \nu_n$ of the elements of any minimal base of Init $\mathcal{A}$ (assuming $\nu_1 \leq \nu_2 \leq \ldots \leq \nu_n$) are independent f the choice of the minimal base, thus these degrees constitute an invariant of Init $\mathcal{A}$ (and consequently, of $\mathcal{A}$). [Hironaka 1], p. 207, denoted this invariant by

$$\nu^*(\mathcal{A}) = \nu^* = (\nu_1, \nu_2, \ldots, \nu_n) .$$

Now let $f_1, f_2, \ldots, f_n \in \mathcal{A}$ be such elements that their initial forms $\overline{f}_1, \overline{f}_2, \ldots, \overline{f}_n$ form a minimal base of Init $\mathcal{A}$. Then by (2.3) $f_1, f_2, \ldots, f_n$ form a base of $\mathcal{A}$.

Suppose, in addition, that $\nu(f_1) \leq \nu(f_2) \leq \ldots \leq \nu(f_n)$; then $\nu(f_i) = \nu_i$.

2.5 Definition ([Hironaka 1], Definition 3 on p. 208). *Such a base $f_1, f_2, \ldots, f_n$ is called a standard base of $\mathcal{A}$.*

2.6 Remark. It may happen that the standard base may not be minimal and the ideal $\mathcal{A}$ may be generated by a subset of the standard base. For example, if $\mathcal{O} = \mathbf{k}[[x, y, z]]$, $\mathcal{A} = (f_1, f_2)$, where $f_1 = x^2y + y^4$, $f_2 = x^3 + z^4$, then the standard base of $\mathcal{A}$ consists of f_1, f_2, f_3 , where $f_3 = xy^4 - yz^4$.

Now recall that we defined a *normal crossing ring* ([Youssin 1], Definition 6.1) as a regular local ring $\mathcal{O}$ with an additional structure consisting of a normal crossing divisor in Spec $\mathcal{O}$ and an ordering of the components of this divisor. We say that a normal crossing ring $\mathcal{O}$ is *complete* if the underlying regular local ring (which, by abuse of notation, we denote by the same letter $\mathcal{O}$) is complete.

We now refer the reader to the definition and discussion of the basic properties of *contact filtrations*; it can be found in Section 9 of [Youssin 1] (in this volume).

2.7 Main Theorem. *Let $\mathcal{O}$ be a complete normal crossing ring of characteristic zero (i.e., $\mathcal{O} \supset \mathbf{Q}$) and let $\mathcal{A} \subset \mathcal{O}$ be an ideal. Consider all contact filtrations Δ in $\mathcal{O}$ with the following property:*

(2.7.1) There exists a standard base $\{f_1, f_2, \ldots, f_n\}$ of $\mathcal{A}$ such that $f_i \in \Delta(\nu_i)$, where $\nu_i = \nu(f_i)$.

Then there exists a minimal filtration Δ_0 among these filtrations, such that it is contained in all the other contact filtrations with this property; of course, Δ_0 is unique.

2.8 Remark. We may reformulate the Main Theorem 2.7 in the following way. For each standard base $\{f_1, f_2, \ldots, f_n\}$ we may consider the contact closure of the filtration generated by $f_1, f_2, \ldots, f_n$ with the weights $\nu_i = \nu(f_i)$, $i = 1, 2, \ldots, n$; then there is a minimal filtration Δ_0 among these contact closures which is contained in all the others.

In Section 9 we shall show that this filtration Δ_0 is the contact closure of the filtration generated by a *normalized* standard base (8.4). However, our approach will be first to construct this filtration Δ_0 and then to show that it contains a standard base (i.e., some standard base $f_1, f_2, \ldots, f_n$ satisfies $f_i \in \Delta_0(\nu_i)$). In a subsequent paper we shall use this approach to extend the construction of the filtration Δ_0 for a class of incomplete rings in which the normalized standard bases might not exist.

We now refer the reader to the definition of *special* filtration and its *first characteristic string* in Section 7 of [Youssin 1] (in this volume).

2.9 Corollary to the Main Theorem 2.7. *In the notation of (2.7) consider all special filtrations* Δ *in* $\mathcal{O}$ *satisfying the property (2.7.1). Then there is a unique filtration* Δ^* *among them whose first characteristic string is lexicographically maximal; any other special filtration with this property has a strictly smaller first characteristic string.*

2.10 Proof: We shall show how to deduce this corollary from the Main Theorem 2.7. Indeed, it is easy to see that the filtration Δ_0 of the Main Theorem 2.7 is finitely Der-generated ([Youssin 1], (14.2)), as it is the minimal contact filtration containing some normalized standard base. Thus we may apply the Main Theorem 7.11 of [Youssin 1] (in its strongest form [Youssin 1], (20.10.1)) to get the special filtration Δ^* associated to Δ_0. Then any other special filtration Δ satisfying the property (2.7.1) contains Δ_0 by (2.7), and the first characteristic string $\operatorname{char}_1 \Delta$ of Δ satisfies $\operatorname{char}_1 \Delta \leq \operatorname{char}_1 \Delta^*$ by the Main Theorem 7.11 of [Youssin 1], and if $\Delta \neq \Delta^*$, then $\operatorname{char}_1 \Delta < \operatorname{char}_1 \Delta^*$. This means that Δ^* is the required filtration. ■

2.11 Remark. This special filtration Δ^* is the canonical polyhedral filtration constructed from the ideal $\mathcal{A}$, that we mentioned in Section 1.

3. Differential operators and principal parts.

From now on we shall assume that all the rings contain the field $\mathbf{Q}$.

3.1 Differential operators. Let $\mathcal{O}$ be a ring and $\mathcal{O}_1 \subset \mathcal{O}$ — a subring. By $\operatorname{Der}_{\mathcal{O}/\mathcal{O}_1}$ we shall denote the $\mathcal{O}$-module of all derivations of $\mathcal{O}$ over $\mathcal{O}_1$ (i.e. vanishing on $\mathcal{O}_1$). For any $\ell \geq 1$ we may consider the module $\operatorname{Diff}^{\ell}_{\mathcal{O}/\mathcal{O}_1}$ of *differential operators* of order ℓ; it is the $\mathcal{O}$-module of all $\mathcal{O}_1$-linear maps $\mathcal{O} \to \mathcal{O}$ which are polynomials of degree $\leq \ell$ in the elements of $\operatorname{Der}_{\mathcal{O}/\mathcal{O}_1}$.

In case $\mathcal{O}_1 = \mathbf{Q}$ we shall denote these modules by $\operatorname{Der}_{\mathcal{O}}$ and $\operatorname{Diff}^{\ell}_{\mathcal{O}}$ respectively.

For any differential operator $D \in \operatorname{Diff}^{\ell}_{\mathcal{O}/\mathcal{O}_1}$ we shall denote its *order* by $\operatorname{ord} D$.

3.2 Principal terms. Let $\mathcal{O}$ be a local ring with the maximal ideal $\mathcal{M}$, $\mathcal{O}/\mathcal{M} = \mathbf{k}$, and, as usual, let

$$\operatorname{gr} \mathcal{O} = \bigoplus_{n=0}^{\infty} \operatorname{gr}_n \mathcal{O} = \bigoplus_{n=0}^{\infty} \mathcal{M}^n/\mathcal{M}^{n+1}.$$

Denote

$$\operatorname{gr}_{\leq \ell} \mathcal{O} = \bigoplus_{n=0}^{\ell} \operatorname{gr}_n \mathcal{O}.$$

Any differential operator $D \in \operatorname{Diff}^{\ell}_{\mathcal{O}}$ maps $\mathcal{M}^{\ell+1}$ into $\mathcal{M}$ and thus it induces a $\mathbf{k}$-linear map $\mathcal{M}^{\ell}/\mathcal{M}^{\ell+1} \to \mathbf{k}$. Thus we get a $\mathcal{O}$-linear map

$$\epsilon_\ell : \operatorname{Diff}^{\ell}_{\mathcal{O}} \to (\operatorname{gr}_\ell \mathcal{O})^* = \operatorname{Hom}(\operatorname{gr}_\ell \mathcal{O}, \mathbf{k}).$$

Clearly, this map factors into a compose

$$\operatorname{Diff}^{\ell}_{\mathcal{O}} \to \operatorname{Diff}^{\ell}_{\mathcal{O}} / \operatorname{Diff}^{\ell-1}_{\mathcal{O}} \to (\operatorname{Diff}^{\ell}_{\mathcal{O}} / \operatorname{Diff}^{\ell-1}_{\mathcal{O}}) \otimes_{\mathcal{O}} \mathbf{k} \to (\operatorname{gr}_\ell \mathcal{O})^* . \tag{3.2.1}$$

From the very definition of the map ϵ_ℓ it follows that if $D \in \operatorname{Diff}^{\ell}_{\mathcal{O}}$ and $f \in \mathcal{M}^{\ell}$, then

$$[(Df) \bmod \mathcal{M}] = \epsilon_\ell(D) \cdot \overline{f} \tag{3.2.2}$$

where

$$\overline{f} = \left[f \bmod \mathcal{M}^{\ell+1}\right] \in \mathcal{M}^{\ell}/\mathcal{M}^{\ell+1} = \mathrm{gr}_{\ell}\,\mathcal{O}$$

is the initial form of f.

More generally, any $D \in \mathrm{Diff}^{\ell}_{\mathcal{O}}$ maps $\mathcal{M}^{\ell+k}$ into $\mathcal{M}^{k}$ for any $k \geq 0$ and thus it induces a map

$$\mathrm{gr}_{\ell+k}\,\mathcal{O} \to \mathrm{gr}_{k}\,\mathcal{O}\,. \tag{3.2.3}$$

On the other hand, $\epsilon_{\ell}(D) \in (\mathrm{gr}_{\ell}\,\mathcal{O})^{*}$ acts as a differential operator of pure order ℓ on $\mathrm{gr}\,\mathcal{O}$; it is easy to see that this action yields the same map (3.2.3).

For any $\mathcal{O}$-submodule $\mathcal{D} \subset \mathrm{Diff}^{\ell}_{\mathcal{O}}$ we define

$$\epsilon(\mathcal{D}) = \bigoplus_{k=0}^{\ell} \epsilon_k(\mathcal{D} \cap \mathrm{Diff}^{k}_{\mathcal{O}}) \subset (\mathrm{gr}_{\leq \ell}\,\mathcal{O})^{*} \tag{3.2.4}$$

which we may consider as the space of principal terms of differential operators from $\mathcal{D}$.

It is easy to see from (3.2.1) that

$$\dim_{\mathbf{k}} \epsilon(\mathcal{D}) \leq \mathrm{rk}_{\mathcal{O}}\,\mathcal{D}\,. \tag{3.2.5}$$

If $D \in \mathrm{Diff}^{\ell}_{\mathcal{O}}$, then we shall also write $\epsilon(D) = \epsilon_{\mathrm{ord}\,D}(D)$.

3.3 Examples. If $\mathcal{O} = \mathbf{k}[[x_1, x_2, \ldots, x_n]]$, then the map ϵ_{ℓ} is onto; the images of the operators

$$\frac{1}{\ell_1!\,\ell_2! \ldots \ell_n!} \cdot \frac{\partial^{\ell}}{\partial x_1^{\ell_1} \partial x_2^{\ell_2} \ldots \partial x_n^{\ell_n}} \tag{3.3.1}$$

where $\sum \ell_i = \ell$, generate the space $(\mathrm{gr}_{\ell}\,\mathcal{O})^{*}$. (More precisely, $\mathrm{gr}_{\ell}\,\mathcal{O}$ is generated by all monomials in $x_1, x_2, \ldots, x_n$ of degree ℓ; the images of the differential operators (3.3.1) form a basis in $(\mathrm{gr}_{\ell}\,\mathcal{O})^{*}$ dual to the basis of monomials in $\mathrm{gr}_{\ell}\,\mathcal{O}$.)

Now suppose that $\mathcal{O}$ is a complete normal crossing ring with the fixed variables u_i, $i \in I$, and $J \subset I$ is any subset. Then, as in [Youssin 1], (9.1.1), we may consider the submodule $\mathrm{Der}_{\mathcal{O},J} \subset \mathrm{Der}_{\mathcal{O}}$ of the derivations preserving the ideals (u_i), $i \in J$:

$$\mathrm{Der}_{\mathcal{O},J} = \left\{ D \in \mathrm{Der}_{\mathcal{O}} \;\middle|\; Du_i \in (u_i) \text{ for } i \in J \right\}.$$

Then we may define the submodule $\mathrm{Diff}^{\ell}_{\mathcal{O},J} \subset \mathrm{Diff}^{\ell}_{\mathcal{O}}$; it consists of those differential operators which are polynomials (of degree $\leq \ell$) in the elements of $\mathrm{Der}_{\mathcal{O},J}$. Then the subspace

$$\epsilon(\mathrm{Diff}^{\ell}_{\mathcal{O},J}) \subset (\mathrm{gr}_{\leq \ell}\,\mathcal{O})^{*}$$

is the orthogonal complement to the subspace

$$\sum_{i \in J} \overline{u}_i \cdot \mathrm{gr}_{\leq \ell - 1}\,\mathcal{O} \subset \mathrm{gr}_{\leq \ell}\,\mathcal{O}\,.$$

Thus $\epsilon(\mathrm{Diff}^{\ell}_{\mathcal{O},J})$ is dual to the space

$$\mathrm{gr}_{\leq \ell}\left[\mathcal{O} \Big/ \sum_{i \in J} (u_i)\right].$$

Another important example is the case of normal crossing rings $\mathcal{O}$ and $\mathcal{O}_1$ such that $\mathcal{O} = \mathcal{O}_1[[x_1, x_2, \ldots, x_n]]$ as normal crossing rings. Then $\mathrm{Diff}^{\ell}_{\mathcal{O}/\mathcal{O}_1} \subset \mathrm{Diff}^{\ell}_{\mathcal{O}}$, and we shall describe $\epsilon(\mathrm{Diff}^{\ell}_{\mathcal{O}/\mathcal{O}_1})$.

Let $\mathcal{M}_1$ be the maximal ideal of $\mathcal{O}_1$, and denote

$$V = \operatorname{Im}(\mathcal{M}_1 \to \mathcal{M}/\mathcal{M}^2) = [\mathcal{M}_1 \bmod \mathcal{M}^2] \subset \operatorname{gr}_1 \mathcal{O} .$$

Then

$$\epsilon(\operatorname{Diff}^{\ell}_{\mathcal{O}/\mathcal{O}_1}) \subset (\operatorname{gr}_{\leq \ell} \mathcal{O})^*$$

is the orthogonal complement to the subspace

$$V \cdot \operatorname{gr}_{\leq \ell - 1} \mathcal{O} \subset \operatorname{gr}_{\leq \ell} \mathcal{O}$$

(i.e., to the intersection with $\operatorname{gr}_{\leq \ell} \mathcal{O}$ of the ideal in $\operatorname{gr} \mathcal{O}$ generated by V). In particular, $\epsilon(\operatorname{Diff}^{\ell}_{\mathcal{O}/\mathcal{O}_1})$ is dual to the space of all polynomials of degree $\leq \ell$ in $x_1, x_2, \ldots, x_n$; the basis in $\epsilon(\operatorname{Diff}^{\ell}_{\mathcal{O}/\mathcal{O}_1})$ dual to the basis of monomials in $x_1, x_2, \ldots, x_n$ is given by the images of the respective monomials in $\partial/\partial x_1, \partial/\partial x_2, \ldots,$ $\partial/\partial x_n$ divided by the suitable factorials (3.3.1).

4. Generalized Fitting ideals.

As we mentioned in Section 1, our way of constructing the filtration Δ_0 of the Main Theorem 2.7 is to differentiate the ideal $\mathcal{A}$ in some delicate way and get the ideals which we call *generalized Fitting ideals*; after that the filtration Δ_0 will be defined as the minimal contact filtration containing these ideals with appropriate weights.

In this section we define these generalized Fitting ideals and prove some of their basic properties. In the next section we shall give some heuristic explanations why these ideals are related to our filtration Δ_0 .

These ideals are related to the determinantal ideals considered by [Fitting], although his context did not involve differential operators. The idea of using the determinantal ideals in the context of differential operators is due to [Villamayor 1] (see also [Villamayor 2]).

4.1 Definition. *Let $\mathcal{O}$ be a ring, $\mathcal{A}$ an ideal in $\mathcal{O}$ and $\mathcal{D}$ a submodule in $\operatorname{Diff}^{\ell}_{\mathcal{O}}$. For any $k \geq 1$ we define the generalized Fitting ideal*

$$\operatorname{FT}^k(\mathcal{D}; \mathcal{A}) \subset \mathcal{O}$$

as the ideal in $\mathcal{O}$ generated by all $k \times k$ determinants $\det(D_i f_j)$, where $D_i \in \mathcal{D}$, $i = 1, 2, \ldots, k$, and $f_j \in \mathcal{A}$, $j = 1, 2, \ldots, k$.

4.2 Remark. Clearly, $\operatorname{FT}^k(\mathcal{D}; \mathcal{A}) \supset \operatorname{FT}^{k+1}(\mathcal{D}; \mathcal{A})$ as any $(k+1) \times (k+1)$ determinant $\det(D_i f_j)$ is a linear combination of $k \times k$ determinants of the same form. Thus $\operatorname{FT}^k(\mathcal{D}; \mathcal{A})$, $k = 1, 2, \ldots$ is a decreasing sequence of ideals in $\mathcal{O}$.

4.3 Definition. *With $\mathcal{O}$ and $\mathcal{A}$ being as in (4.1), we define $k_*(\mathcal{D}; \mathcal{A})$ to be the smallest k such that $\operatorname{FT}^k(\mathcal{D}; \mathcal{A}) \neq (1)$; if such k does not exist, we shall say $k_*(\mathcal{D}; \mathcal{A})$ is undefined.* We define*

$$\operatorname{FT}^*(\mathcal{D}; \mathcal{A}) = \operatorname{FT}^{k_*(\mathcal{D}; \mathcal{A})}(\mathcal{D}; \mathcal{A}) .$$

4.4 Proposition. *Let $\mathcal{O}_1 \subset \mathcal{O}$ be a subring, and assume $\mathcal{D} \subset \operatorname{Diff}^{\ell}_{\mathcal{O}/\mathcal{O}_1}$. Pick up a base D_α , $\alpha \in A$, of $\mathcal{D}$ as an $\mathcal{O}$-module, and a base f_β , $\beta \in B$, of $\mathcal{A}$ as an $\mathcal{O}_1$-module (both bases need not be*

* This will not be the case in any of the examples in which we are interested.

finite or even countable). Then the ideal $\mathrm{FT}^k(\mathcal{D};\mathcal{A})$ *is generated by all* $k\times k$ *determinants* $\det(D_{\alpha_i} f_{\beta_j})$, *where* $\alpha_1,\alpha_2,\dots,\alpha_k\in A$ *and* $\beta_1,\beta_2,\dots,\beta_k\in B$.

The proof is obvious. ■

Unfortunately, in an important case $\mathcal{O}=\mathcal{O}_1[[x_1,x_2,\dots,x_n]]$ the base of $\mathcal{A}$ as an $\mathcal{O}_1$-module may well be uncountable and far from being easy to describe, so Proposition 4.4 may not be helpful in computing the ideals $\mathrm{FT}^k(\mathcal{D};\mathcal{A})$. The following result is useful in this case.

4.5 Proposition. *In the notation of (4.4) suppose that* $\mathcal{O}$ *is a local ring with a maximal ideal* $\mathcal{M}$. *Take a base* D_α, $\alpha\in A$, *of* $\mathcal{D}$ *as an* $\mathcal{O}$*-module, and a topological base* f_β, $\beta\in B$ *of* $\mathcal{A}$ *as an* $\mathcal{O}_1$*-module (* $\mathcal{O}$ *is considered as having* $\mathcal{M}$*-adic topology). Then the ideal* $\mathrm{FT}^k(\mathcal{D};\mathcal{A})$ *is generated by all* $k\times k$ *determinants* $\det(D_{\alpha_i} f_{\beta_j})$ *where* $\alpha_1,\alpha_2,\dots,\alpha_k\in A$ *and* $\beta_1,\beta_2,\dots,\beta_k\in B$.

The proof follows easily from the closedness of the ideal $\mathrm{FT}^k(\mathcal{D};\mathcal{A})$ in $\mathcal{M}$-adic topology in $\mathcal{O}$. ■

We conclude this section with the following general result on the number $k_*(\mathcal{D};\mathcal{A})$.

4.6 Proposition. *If* $\mathcal{O}$ *is a local ring with maximal ideal* $\mathcal{M}$ *and residue field* $\mathbf{k}$, *and* $\mathcal{D}$ *is a submodule in* $\mathrm{Diff}^\ell_{\mathcal{O}}$, *then* $k_*(\mathcal{D},\mathcal{A})$ *is strictly greater than the rank of the pairing*

$$\epsilon(\mathcal{D})\times \mathrm{Init}_{\le\ell}\mathcal{A}\to\mathbf{k}. \tag{4.6.1}$$

(The pairing is the restriction of the pairing of $(\mathrm{gr}\,\mathcal{O})^*$ *with* $\mathrm{gr}\,\mathcal{O}$ *to* $\epsilon(\mathcal{D})\subset(\mathrm{gr}\,\mathcal{O})^*$ *and* $\mathrm{Init}_{\le\ell}\mathcal{A}\subset\mathrm{gr}\,\mathcal{O}$ *.)*

4.7 Proof: Let r be the rank of the pairing (4.6.1). Then we can find $D_1,D_2,\dots,D_r\in\mathcal{D}$, $\mathrm{ord}\,D_i=\ell_i$, and $f_1,f_2,\dots,f_r\in\mathcal{A}$, $\nu(f_j)=\ell_j$, such that

$$\det\bigl(\epsilon(D_i)\cdot\overline{f}_j\bigr)\neq 0$$

(Here $\overline{f}_j=\bigl[f_j \bmod \mathcal{M}^{\ell_j+1}\bigr]$ is the initial form of f_j.)

By (3.2.2)

$$\epsilon(D_i)\cdot\overline{f}_j=\bigl[(D_if_j)\bmod\mathcal{M}\bigr]$$

if $\ell_i\le\ell_j$, and

$$\epsilon(D_i)\cdot\overline{f}_j=\bigl[(D_if_j)\bmod\mathcal{M}\bigr]=0$$

if $\ell_i<\ell_j$; thus the matrix $(D_if_j \bmod \mathcal{M})$ is block-triangular (blocks correspond to the groups of D_i and f_j with the same value of ℓ_i and ℓ_j), and the diagonal blocks are the same as the diagonal blocks of the matrix $(\epsilon(D_i)\cdot\overline{f}_j)$. However, the latter matrix does not have any nonzero blocks outside the diagonal, thus

$$\det(D_if_j \bmod \mathcal{M})=\det(\epsilon(D_i)\cdot\overline{f}_j)\neq 0$$

so

$$\det(D_if_j)\notin\mathcal{M}.$$

This shows that

$$\mathrm{FT}^r(\mathcal{D};\mathcal{A})=(1)$$

and consequently

$$r<k_*(\mathcal{D};\mathcal{A}).$$

■

5. Heuristics.

Here we present some heuristic explanations of how the generalized Fitting ideals are related to the filtration Δ_0 of the Main Theorem 2.7.

We first explain how the generalized Fitting ideals are related to the Hilbert-Samuel stratum.

5.1 Remark (Pierre Milman, private communication; see also [Bierstone, Milman 3]). We shall show that the closed subset defined by the ideal

$$\mathbf{b} = \mathcal{A} + \sum_{\ell=1}^{\infty} \mathrm{FT}^*(\mathrm{Diff}^{\ell}_{\mathcal{O}}; \mathcal{A}) \tag{5.1.1}$$

is the stratum where the Hilbert-Samuel function of the scheme $\mathrm{Spec}(\mathcal{O}/\mathcal{A})$ is maximal.

We shall have to make some assumptions on the ring $\mathcal{O}$; we shall assume that $X = \mathrm{Spec}\,\mathcal{O}$ is a smooth algebraic variety over a field $\mathbf{k}$.

Consider the following Hilbert-Samuel function of $\mathrm{Spec}(\mathcal{O}/\mathcal{A})$ at a point $x \in X$; it is a function of the integer argument ℓ defined by

$$H(\mathrm{Spec}(\mathcal{O}/\mathcal{A}); x; \ell) = \mathrm{length}\big(\mathcal{O}_x/(\mathcal{A} + \mathcal{M}_x^{\ell+1})\big) \tag{5.1.2}$$

where $\mathcal{M}_x \subset \mathcal{O}$ is the ideal of the point x . (Note that this Hilbert-Samuel function is different from the one in Section 6.)

The Hilbert-Samuel function defines a stratification of X ; each stratum is the set of points $x \in X$ where the Hilbert-Samuel function has a given value for each ℓ .

The Hilbert-Samuel function is upper-semicontinuous ([Bennett], Theorem (2)). Hence, the innermost Hilbert-Samuel stratum is the set of points $x \in X$ where the Hilbert-Samuel function is maximal (i.e., (5.1.2) is maximal for all ℓ).

Now consider the module of ℓ-th differentials of the ring $\mathcal{O}$ over $\mathbf{k}$:

$$\widetilde{\mathrm{Diff}}^{\ell}_{\mathcal{O}} = \mathcal{P}_{X/\mathbf{k}} \big/ \mathcal{P}^{\ell+1}_{X/\mathbf{k}}$$

where $\mathcal{P}_{X/\mathbf{k}}$ is the ideal of the diagonal $\Delta \subset X \times_{\mathbf{k}} X$ (see [EGA $\mathrm{IV_{IV}}$], 16.3).

$\widetilde{\mathrm{Diff}}^{\ell}_{\mathcal{O}}$ becomes an $\mathcal{O}$-module by one of the projections $X \times_{\mathbf{k}} X \to X$, and the other projection yields a map

$$d^{\ell} : \mathcal{O} \to \widetilde{\mathrm{Diff}}^{\ell}_{\mathcal{O}}$$

which we may consider "the full differential" (it is not $\mathcal{O}$-linear). Any differential operator $D \in \mathrm{Diff}^{\ell}_{\mathcal{O}}$, $D : \mathcal{O} \to \mathcal{O}$, is a composition of d^{ℓ} and some $\mathcal{O}$-linear map $\widetilde{\mathrm{Diff}}^{\ell}_{\mathcal{O}} \to \mathcal{O}$. Thus differential operators correspond to the linear maps

$$\widetilde{\mathrm{Diff}}^{\ell}_{\mathcal{O}} \to \mathcal{O}$$

and the $\mathcal{O}$-modules $\widetilde{\mathrm{Diff}}^{\ell}_{\mathcal{O}}$ and $\mathrm{Diff}^{\ell}_{\mathcal{O}}$ are dual.

It is easy to see that there is a $\mathbf{k}$-linear isomorphism

$$\mathcal{O}\big/(\mathcal{A} + \mathcal{M}_x^{\ell}) \cong \frac{\widetilde{\mathrm{Diff}}^{\ell}_{\mathcal{O}}}{(d^{\ell}\mathcal{A})} \otimes_{\mathcal{O}} (\mathcal{O}/\mathcal{M}_x) \, .$$

Hence, the points where the Hilbert-Samuel function is maximal, are the points x where the map

$$(d^\ell \mathcal{A}) \otimes_{\mathcal{O}} \mathcal{O}/\mathcal{M}_x \to \widetilde{\mathrm{Diff}}^\ell_{\mathcal{O}} \otimes_{\mathcal{O}} (\mathcal{O}/\mathcal{M}_x)$$

has minimal rank. (Here $(d^\ell \mathcal{A})$ is the $\mathcal{O}$-submodule of $\widetilde{\mathrm{Diff}}^\ell_{\mathcal{O}}$ generated by $d^\ell \mathcal{A}$.) It is not hard to see that this set is the set of zeroes of the Fitting ideal $\mathrm{FT}^*(\mathrm{Diff}^\ell_{\mathcal{O}}; \mathcal{A})$ (and then the value of the minimal rank is $k_*(\mathrm{Diff}^\ell_{\mathcal{O}}; \mathcal{A}) - 1$). Thus the set of common zeroes of the Fitting ideals $\mathrm{FT}^*(\mathrm{Diff}^\ell_{\mathcal{O}}; \mathcal{A})$ for all ℓ coincides with the Hilbert-Samuel stratum of the variety $\mathrm{Spec}(\mathcal{O}/\mathcal{A})$.

5.2 Remark. Now we shall show how the Hilbert-Samuel strata are related to our filtration Δ_0. Suppose $\mathcal{O}$ is a complete regular local ring. The Hilbert-Samuel strata are the strata where the symbol $\nu_*(\mathcal{A})$ (see (2.4)) is constant.* If, in addition, we can find a standard base $\{f_1, f_2, \ldots, f_n\}$ of $\mathcal{A}$, $\nu(f_i) = \nu_i$, $\nu_*(\mathcal{A}) = (\nu_1, \nu_2, \ldots, \nu_n)$, such that it is a standard base at any (not necessarily closed) point of the same Hilbert-Samuel stratum, then this stratum is the set of points where $\nu(f_i) = \nu_i$. (A *normalized* standard base has this property locally in a small neighborhood, cf. [Bierstone, Milman 2], Theorem 5.3.1.)

Thus, the Hilbert-Samuel stratum is the zero set of the ideal

$$\mathbf{b}' = \sum_{i=1}^{n} \mathrm{Diff}^{\nu_i - 1}_{\mathcal{O}} f_i \tag{5.2.1}$$

(cf. [Giraud 1], (2.3.3)). Hence, the ideal $\mathbf{b}'$ (5.2.1) should be closely related to the ideal $\mathbf{b}$ (5.1.1); at least, they have a common zero set. In other words, $\mathbf{b}'$ is somehow related to the ideals $\mathrm{FT}^*(\mathcal{D}, \mathcal{A})$.

The definitions of the filtration Δ_0 and the ideal $\mathbf{b}'$ are similar in the following way. Δ_0 as a contact filtration is generated by the elements of some standard base, with higher elements being assigned higher weights; $\mathbf{b}'$ is defined by differentiating the elements of some standard base with higher elements being differentiated more times.

Thus we see that there is a relationship between the ideal $\mathbf{b}'$ and the Fitting ideals $\mathrm{FT}^*(\mathcal{D}; \mathcal{A})$, and the filtration Δ_0 is defined in a way similar to the way the ideal $\mathbf{b}'$ is defined. This suggests that there could be a relationship between the filtration Δ_0 and the ideals $\mathrm{FT}^*(\mathcal{D}; \mathcal{A})$. More precisely, in Sections 7–9 we shall show that Δ_0 is generated by the ideals $\mathrm{FT}^*(\mathcal{D}; \mathcal{A})$ for certain submodules $\mathcal{D} \subset \mathrm{Diff}^\ell_{\mathcal{O}}$.

Note that the ideal $\mathbf{b}$ is defined canonically by the ideal $\mathcal{A}$; at the same time the ideal $\mathbf{b}'$ depends on the choice of the standard base f_i, and the higher elements of the standard base are differentiated more times. This shows that the Fitting ideals in some sense allow us to "get hold of" the higher elements of the standard base.

The following result also illustrates the relationship between the ideals $\mathbf{b}$ (5.1.1) and $\mathbf{b}'$ (5.2.1).

5.3 Proposition. *If $\mathcal{O}$ is a complete regular local ring with the maximal ideal $\mathcal{M}$, then*

$$(\mathbf{b} \bmod \mathcal{M}^2) = (\mathbf{b}' \bmod \mathcal{M}^2) .$$

5.4 Remark. It is easy to see that $\mathbf{b}' \bmod \mathcal{M}^2$ is the minimal subspace in $\mathcal{M}/\mathcal{M}^2$ such that the ideal $\mathrm{Init}\,\mathcal{A} \subset \mathrm{gr}\,\mathcal{O}$ can be generated by polynomials in the elements of this subspace.

5.5 Sketch of the proof of (5.3) : For any ℓ, find $g_1, g_2, \ldots, g_k \subset \mathcal{A}$ and $D_1, D_2, \ldots, D_N \in \mathrm{Diff}^\ell_{\mathcal{O}}$ with the following properties:

---

* This fact can be proven by the same method that [Bennett] applied to prove his Theorem (2); Cor. 1 on p. 196 of [Hironaka 1] and Theorems I, II, III of [Hironaka 3] have to be used.

(i) The initial forms of $g_1, g_2, \ldots, g_k$ form a basis of of the linear space $\mathrm{Init}_{\leq \ell} \mathcal{A}$ and the principal terms of $D_1, D_2, \ldots, D_N$ form a basis of the space $(\mathrm{gr}_{\leq \ell} \mathcal{O})^*$.

(ii) The restrictions of the principal terms of $D_1, D_2, \ldots, D_k$ onto $\mathrm{Init}_{\leq \ell} \mathcal{A}$ (as linear functionals on $\mathrm{gr}_{\leq \ell} \mathcal{O}$) form a basis of $\left(\mathrm{Init}_{\leq \ell} \mathcal{A}\right)^*$ dual to the initial forms of $g_1, g_2, \ldots, g_k$.

(iii) If $\mathrm{ord}\, D_i \geq \nu(g_j)$ (here $1 \leq i \leq N$, $1 \leq j \leq k$), then $D_i g_j = \delta_{ij}$.

(It is not hard to see that we can find such $g_1, g_2, \ldots, g_k$, $D_1, D_2, \ldots, D_N$.)

Then for any k' , $\mathrm{FT}^{k'}(\mathrm{Diff}^{\ell}_{\mathcal{O}}; \mathcal{A})$ is generated by all the determinants

$$\det(D'_i g'_j) ,$$

where each D'_i , $1 \leq i \leq k'$, is one of $D_1, D_2, \ldots, D_N$, and each g'_j , $1 \leq j \leq k'$, is either one of $g_1, g_2, \ldots, g_k$, or $g'_j \in \mathcal{A} \cap \mathcal{M}^{\ell+1}$.

Clearly, $\mathrm{FT}^{k'}(\mathrm{Diff}^{\ell}_{\mathcal{O}}; \mathcal{A}) = (1)$ for $k' \leq k$, and

$$\mathrm{FT}^{k+1}(\mathrm{Diff}^{\ell}_{\mathcal{O}}; \mathcal{A}) \neq (1) .$$

Moreover, the subspace

$$\left[\mathrm{FT}^{k+1}(\mathrm{Diff}^{\ell}_{\mathcal{O}}; \mathcal{A}) \bmod \mathcal{M}^2\right] \subset \mathcal{M}/\mathcal{M}^2$$

is generated by all the elements

$$[Dg \bmod \mathcal{M}^2] \in \mathcal{M}/\mathcal{M}^2 ,$$

where $\mathrm{ord}\, D = \ell$, $\nu(g) = \ell + 1$, $g \in \mathcal{A}$ and $\epsilon_\ell(D)$ is orthogonal to the subspace $\mathrm{Init}_\ell(\mathcal{A})$.

Finally, it is not hard to see that the sum of these subspaces for all $\ell \geq 1$ coincides with the minimal subspace of $\mathcal{M}/\mathcal{M}^2$ such that the ideal $\mathrm{Init}\, \mathcal{A}$ can be generated by the elements of this subspace. ■

6. Generic position.

Let $\mathcal{O}$ be a regular local ring with the maximal ideal $\mathcal{M}$, $\mathcal{A}$ an ideal in $\mathcal{O}$ and $\mathcal{O}_1$ a subring of $\mathcal{O}$ with the maximal ideal $\mathcal{M}_1$, such that $\mathcal{O}/\mathcal{M} = \mathcal{O}_1/\mathcal{M}_1 = \mathbf{k}$, where $\mathbf{k}$ is some field. We shall assume that both $\mathcal{O}_1$ and $\mathcal{O}/\mathcal{M}_1\mathcal{O}$ are regular local rings (in other words, $\mathcal{O}_1$ is regular and the regular system of parameters in $\mathcal{O}_1$ can be extended to a regular system of parameters in $\mathcal{O}$).

In this section we study the following question: what does it mean that the ideal $\mathcal{A}$ and the subring $\mathcal{O}_1$ are in generic position?

The property of generic position we are interested in is, in fact, a property of the ideal $\mathrm{Init}\, \mathcal{A} \subset \mathrm{gr}\, \mathcal{O}$ and the subring $\mathrm{gr}\, \mathcal{O}_1 \subset \mathrm{gr}\, \mathcal{O}$ being in generic position.

Let $V = \mathcal{M}/\mathcal{M}^2$ and $V_1 = \mathcal{M}_1/\mathcal{M}_1^2$; then $V_1 \subset V$, $\mathrm{gr}\, \mathcal{O} = \mathbf{k}[V]$, $\mathrm{gr}\, \mathcal{O}_1 = \mathbf{k}[V_1]$. Let $A = \mathrm{Init}\, \mathcal{A}$; it is a homogeneous ideal in $\mathbf{k}[V]$. We are interested in the property of the ideal $A \subset \mathbf{k}[V]$ and the subring $\mathbf{k}[V_1] \subset \mathbf{k}[V]$ being in generic position.

6.1 Hilbert-Samuel functions. We change our terminology now; we shall call *the Hilbert-Samuel function* not the function (5.1.2), but rather the following related object.

Let R be a local ring with a maximal ideal $\mathcal{M}_R$. We shall say that its *Hilbert-Samuel function* is given by

$$H(R; n) = \dim(\mathcal{M}_R^n/\mathcal{M}_R^{n+1}) . \tag{6.1.1}$$

Clearly, $H(R) = H(\operatorname{gr} R)$.

If $H(n)$ is a function on the set $\mathbf{Z}_+$ of nonnegative integers, then we may define the operators

(6.1.2) $$(\nabla H)(n) = H(n) - H(n-1)$$

(assuming $H(-1) = 0$) , and

(6.1.3) $$(\nabla^{-1} H)(n) = \sum_{i=0}^{n} H(i) .$$

Clearly, the operators ∇ and ∇^{-1} are inverse to each other, and R is a regular local ring of dimension d if and only if

$$H(R) = \nabla^{-d} H(\mathbf{k}) ,$$

where $H(\mathbf{k})$ is the Hilbert-Samuel function of a field:

(6.1.4) $$H(\mathbf{k}; n) = \begin{cases} 1, & \text{if } n = 0; \\ 0, & \text{if } n > 0. \end{cases}$$

6.2 Proposition ([Hironaka 3], Proposition 6; [Giraud 2], Lemme I.3.9; [Giraud 3], Lemme 1.2; [Singh], Theorem 1). *The following properties of the triple* $\mathbf{k}[V]$, $\mathbf{k}[V_1]$, A *(as above) are equivalent:*

(6.2.1) Any basis $x_1, x_2, \ldots, x_d$ *of* V_1 *is a regular sequence in* $\mathbf{k}[V]/A$ *(this means that each* x_i *is not a zerodivisor in* $\mathbf{k}[V]/(x_1, x_2, \ldots, x_{i-1}, A)$ *).*

(6.2.2) Let $\mathbf{k}[V]_{(V)}$ *be the localization of* $\mathbf{k}[V]$ *at the maximal ideal generated by* V *, and let* $d = \dim V_1$ *; then*

$$H\Big(\frac{\mathbf{k}[V]_{(V)}}{(V_1, A)}\Big) = \nabla^d H\big(\mathbf{k}[V]_{(V)}/A\big) .$$

(6.2.3) Consider any base $\overline{f}_\alpha$, $\alpha \in S$ *, of the ideal*

$$[A \bmod V_1] \subset \mathbf{k}[V]/(V_1)$$

considered as a vector space over $\mathbf{k}$ *. If* f_α , $\alpha \in S$ *, is any lift of* $\overline{f}_\alpha$ *to* $\mathbf{k}[V]$ *, such that* $f_\alpha \in A$ *, then* f_α , $\alpha \in S$ *is a base of* A *as a* $\mathbf{k}[V_1]$*-module.*

6.3 Proof: The equivalence of (6.2.1) and (6.2.3) is an easy exercise which is left to the reader.

Now let us show the equivalence of (6.2.1) and (6.2.2). Clearly, it is enough to consider only the case $d = \dim V_1 = 1$ as the general case follows by induction. Let x be a nonzero element of V_1 , and let

$$R = \mathbf{k}[V]_{(V)}/A$$
$$\mathcal{M}_R = V \cdot R .$$

The condition (6.2.1) means that x is not a zerodivisor in R , or equivalently, that the map

$$\mathcal{M}_R^{n-1}/\mathcal{M}_R^n \xrightarrow{\times x} \mathcal{M}_R^n/\mathcal{M}_R^{n+1}$$

is an injection for all n . This, however, is equivalent to

$$\dim \frac{\mathcal{M}_R^n}{(x) \cap \mathcal{M}_R^n + \mathcal{M}_R^{n+1}} = \dim \mathcal{M}_R^n/\mathcal{M}_R^{n+1} - \dim \mathcal{M}_R^{n-1}/\mathcal{M}_R^n$$

or to

$$H(R/(x)) = \nabla H(R)$$

and this is the condition (6.2.2) in our case. ∎

6.4 Remark. Property (6.2.2) (which is a property of the triple $\mathbf{k}[V], \mathbf{k}[V_1], A$) says that

$$H\Big(\frac{\mathbf{k}[V]_{(V)}}{(V_1, A)}\Big) = \nabla^d H\big(\mathbf{k}[V]_{(V)}/A\big) .$$

In general, this property does not have to be satisfied; the following inequality is true in general:

$$\nabla^{-d+1} H\Big(\frac{\mathbf{k}[V]_{(V)}}{(V_1, A)}; n\Big) \geq \nabla H\big(\mathbf{k}[V]_{(V)}/A; n\big) . \tag{6.4.1}$$

Clearly, the equality in (6.4.1) holds if and only if the condition (6.2.2) is satisfied; to see this, one has to apply the operator ∇^{-d+1} (which is an isomorphism) to both sides of (6.2.2).

It is easy to see that the reasoning of (6.3) yields the inequality (6.4.1). This inequality is related to more general results of [Giraud 2, 3], [Singh], [Hironaka 3] and [Bennett]. Namely, their results are good for any, not necessarily graded local ring; the results of [Giraud 2], Lemme I.3.9, [Giraud 3], Lemme 1.2, and [Singh] yield the following corollary from (6.4.1) :

$$\nabla^{-d} H\Big(\frac{\mathbf{k}[V]_{(V)}}{(V_1, A)}; n\Big) \geq H\big(\mathbf{k}[V]_{(V)}/A; n\big)$$

The results of [Bennett], Theorem (1) and [Hironaka 3], Proposition 5 yield that

$$\nabla^{-d-1} H\Big(\frac{\mathbf{k}[V]_{(V)}}{(V_1, A)}; n\Big) \geq \nabla^{-1} H\big(\mathbf{k}[V]_{(V)}/A; n\big)$$

Both these inequalities follow from (6.4.1) by an application of the operator ∇^{-1} which is clearly monotonous on the functions $\mathbf{Z}_+ \to \mathbf{Z}_+$.

6.5 Definition. *Let $\mathcal{O}$, $\mathcal{O}_1$ and $\mathcal{A}$ be as above. We shall say that the subring $\mathcal{O}_1$ is in generic position to the ideal $\mathcal{A}$ if the triple* $\operatorname{gr}\mathcal{O}$, $\operatorname{gr}\mathcal{O}_1$ *and* $\operatorname{Init}\mathcal{A}$ *satisfies the conditions of Proposition 6.2.*

6.6 Remark. The reason we use the term "generic position" here, is as follows. Let the notation be as above, and suppose the ring $\mathcal{O}$ and the ideal $\mathcal{A} \subset \mathcal{O}$ are fixed, and let us see what kind of restriction the property of being in generic position imposes on the subring $\mathcal{O}_1$. First of all, from (6.2.1) we see that

$$d = \dim \mathcal{O}_1 = \dim V_1 \leq \operatorname{depth} \mathbf{k}[V]/A \tag{6.6.1}$$

Now suppose we have fixed some value d satisfying (6.6.1); then it is not hard to see that (6.2.1) is a Zarisky-open condition on the subspace V_1 .

In other words, the restriction on the choice of the subring $\mathcal{O}_1 \subset \mathcal{O}$ is that, first, its dimension is small enough, and second, its "direction" V_1 satisfies some Zarisky-open condition; this is, indeed, a generic position restriction.

Note that the property of $\mathcal{O}_1$ to be in generic position to $\mathcal{A}$ is equivalent to the map $\operatorname{Spec}(\mathcal{O}/\mathcal{A}) \to \operatorname{Spec}\mathcal{O}_1$ being *transverse* to $\operatorname{Spec}(\mathcal{O}/\mathcal{A})$ in the sense of [Giraud 1], Definition 3.1.

6.7 Example. Let $W \subset \mathcal{M}/\mathcal{M}^2 = V$ be the minimal subspace such that the ideal $\operatorname{Init}\mathcal{A} \subset \mathbf{k}[V] = \operatorname{gr}\mathcal{O}$ can be generated by polynomials in the elements of W . If the subspaces

$$V_1 = \mathcal{M}_1/\mathcal{M}_1^2 \subset V$$

and

$$W \subset V$$

do not have a nonzero intersection, then it is easy to see that (6.2.1) is satisfied, so $\mathcal{O}_1$ is in generic position to $\mathcal{O}$. In particular, this is the case when $x_1, x_2, \ldots, x_n \in \mathcal{M}$ are such that $\operatorname{Init} \mathcal{A}$ can be generated by polynomials in $x_1, x_2, \ldots, x_m$ and

$$\mathcal{O} = \mathcal{O}_1[[x_1, x_2, \ldots, x_N]] .$$

However, it is possible that $\mathcal{O}_1$ is in generic position to $\mathcal{A}$ even when $\operatorname{Init} \mathcal{A}$ cannot be generated by elements of some subspace complementary to $\mathcal{O}_1$.

7. Fitting ideals and filtrations generated by standard bases.

Let $\mathcal{O}$ be a complete normal crossing ring, $\mathcal{O} \supset \mathbf{Q}$, and let $\mathcal{A} \subset \mathcal{O}$ be an ideal. Let Δ, $\Delta \neq (1)$, be a contact filtration in $\mathcal{O}$ containing some standard base of $\mathcal{A}$ (i.e., some standard base $\{f_1, f_2, \ldots, f_n\}$ of $\mathcal{A}$ satisfies $f_i \in \Delta(\nu(f_i))$). In this section we make the first step towards the proof of the Main Theorem 2.7 and construct some Fitting ideals contained in the filtration Δ ; in (7.10) we define the filtration Δ_0 which, as we shall show in Section 9, satisfies the properties of the Main Theorem 2.7.

7.1 Notation. Let $\{f_1, f_2, \ldots, f_n\}$ be a standard base of $\mathcal{A}$ such that

(7.1.1) $f_i \in \Delta(\nu_i)$, where $\nu_i = \nu(f_i)$.

Let $\mathcal{M}$ be the maximal ideal of $\mathcal{O}$, $\mathbf{k} = \mathcal{O}/\mathcal{M}$, and let $W \subset \mathcal{M}/\mathcal{M}^2$ be the minimal subspace which is weakly transverse to the fixed variables (i.e., it is transverse to all those fixed variables which are not contained in it) and such that the ideal $\operatorname{Init} \mathcal{A}$ can be generated by elements of $\mathbf{k}[W]$ (see [Youssin 1], Section 8).

We shall consider a normal crossing subring $\mathcal{O}_1 \subset \mathcal{O}$ and elements $x_1, x_2, \ldots, x_m \in \mathcal{M}$ having the following properties:

(7.1.2) $\mathcal{O} = \mathcal{O}_1[[x_1, x_2, \ldots, x_m]]$ as normal crossing rings;

(7.1.3) $\mathcal{O}_1$ is in generic position to $\mathcal{A}$;

(7.1.4) $x_1, x_2, \ldots, x_m \in \Delta(1)$;

(7.1.5) $[x_i \bmod \mathcal{M}^2] \in W$, $i = 1, 2, \ldots, m$.

It is not hard to see that a presentation

$$\mathcal{O} = \mathcal{O}_1[[x_1, x_2, \ldots, x_m]]$$

satisfying these properties, does exist. Indeed, (7.1.1) yields

$$W \subset [\Delta(1) \bmod \mathcal{M}^2]$$

so we may take such $x_1, x_2, \ldots, x_m \in \Delta(1)$ that their images in $\mathcal{M}/\mathcal{M}^2$ form a base of $W \subset \mathcal{M}/\mathcal{M}^2$, and $\mathcal{O}_1$ such that $\mathcal{O} = \mathcal{O}_1[[x_1, x_2, \ldots, x_m]]$. Then by (6.7) $\mathcal{O}_1$ is in generic position to $\mathcal{A}$, and the properties (7.1.2)-(7.1.5) are satisfied.

7.2 Remark. Consider a submodule $\mathcal{D} \subset \operatorname{Diff}^d_{\mathcal{O}/\mathcal{O}_1}$, where d is some positive integer. As we noted in (3.3),

$$\epsilon(\operatorname{Diff}^d_{\mathcal{O}/\mathcal{O}_1}) \subset (\mathbf{k}[\overline{x}_1, \overline{x}_2, \ldots, \overline{x}_m])^*$$

where $\overline{x}_i = [x_i \bmod \mathcal{M}^2]$; thus also

$$\epsilon(\mathcal{D}) \subset (\mathbf{k}[\overline{x}_1, \overline{x}_2, \ldots, \overline{x}_m])^* .$$

On the other hand, let $\mathcal{M}_1$ be the maximal ideal in $\mathcal{O}_1$, let $V_1 = \mathcal{M}_1/\mathcal{M}_1^2 \subset \mathcal{M}/\mathcal{M}^2 = \mathrm{gr}_1\,\mathcal{O}$, and let (V_1) be the ideal in $\mathrm{gr}\,\mathcal{O}$ generated by V_1. Then clearly

$$\frac{\mathrm{gr}\,\mathcal{O}}{(V_1)} \simeq \mathbf{k}[\overline{x}_1, \overline{x}_2, \ldots, \overline{x}_m]$$

In particular,

$$[(\mathrm{Init}\,\mathcal{A}) \bmod (V_1)] \subset \mathbf{k}[\overline{x}_1, \overline{x}_2, \ldots, \overline{x}_m]$$

and we get a pairing between $\epsilon(\mathcal{D})$ and $(\mathrm{Init}\,\mathcal{A}) \bmod (V_1)$.

Note also that by (3.2.5)

$$\dim \epsilon(\mathcal{D}) \leq \mathrm{rk}_{\mathcal{O}}\,\mathcal{D}\,.$$

7.3 Definition. *We shall say that a submodule $\mathcal{D} \subset \mathrm{Diff}^d_{\mathcal{O}/\mathcal{O}_1}$ is d-dual to $\mathcal{A}$ if the following conditions are satisfied:*

(7.3.1) $\dim \epsilon(\mathcal{D}) = \mathrm{rk}_{\mathcal{O}}\,\mathcal{D}$.

(7.3.2) The pairing between $\epsilon(\mathcal{D})$ and $(\mathrm{Init}_{\leq d}\,\mathcal{A}) \bmod (V_1)$ is nondegenerate.

(In particular, this means that $\dim \epsilon(\mathcal{D}) = \mathrm{rk}_{\mathcal{O}}\,\mathcal{D} = \dim[(\mathrm{Init}_{\leq d}\,\mathcal{A}) \bmod (V_1)]$.)

The following proposition is the main result of this section.

7.4 Proposition. *Consider a submodule $\mathcal{D} + \mathcal{O} \subset \mathrm{Diff}^d_{\mathcal{O}/\mathcal{O}_1}$ where $\mathcal{D} \subset \mathrm{Diff}^d_{\mathcal{O}/\mathcal{O}_1}$ is d-dual to $\mathcal{A}$ and $\mathcal{O}$ acts as differential operators of order zero. Then*

$$k_*(\mathcal{D} + \mathcal{O}; \mathcal{A}) = \dim \epsilon(\mathcal{D}) + 1$$

$$\mathrm{FT}^*(\mathcal{D} + \mathcal{O}; \mathcal{A}) \subset \Delta(d+1)$$

To prove this proposition, we first need a lemma.

7.5 Lemma. *There exists a topological (in $\mathcal{M}$-adic topology) base $g_1, g_2, \ldots$ of $\mathcal{A}$ as an $\mathcal{O}_1$-module, with the following properties:*

(7.5.1) Let $\overline{g}_i = [g_i \bmod \mathcal{M}^{\nu(g_i)+1}] \in \mathrm{gr}_{\nu(g_i)}\,\mathcal{O}$ be the initial form of g_i, and let $\widehat{g}_i = [\overline{g}_i \bmod (V_1)] \in (\mathrm{gr}\,\mathcal{O})/(V_1)$; then $\widehat{g}_1, \widehat{g}_2, \ldots$ form a basis of $(\mathrm{Init}\,\mathcal{A}) \bmod (V_1)$ as a linear space over $\mathbf{k}$.

(7.5.2) Let $N = \dim \epsilon(\mathcal{D}) = \mathrm{rk}_{\mathcal{O}}\,\mathcal{D} = \dim[(\mathrm{Init}_{\leq d}\,\mathcal{A}) \bmod (V_1)]$ (all these numbers are equal as $\mathcal{D}$ is d-dual to $\mathcal{A}$). Then $\widehat{g}_1, \widehat{g}_2, \ldots, \widehat{g}_N$ form a basis of $(\mathrm{Init}_{\leq d}\,\mathcal{A}) \bmod (V_1)$.

(7.5.3) $g_i \in \Delta(\nu(g_i))$.

(7.6) Proof: Indeed, if we take any basis $\{\widehat{g}_1, \widehat{g}_2, \ldots\}$ of $(\mathrm{Init}\,\mathcal{A}) \bmod (V_1)$ (as a linear space over $\mathbf{k}$) such that $\widehat{g}_1, \widehat{g}_2, \ldots, \widehat{g}_N$ form a basis of $(\mathrm{Init}_{\leq d}\,\mathcal{A}) \bmod (V_1)$, and lift it first to $\overline{g}_1, \overline{g}_2, \ldots \in \mathrm{Init}\,\mathcal{A}$ and then to $g_1, g_2, \ldots \in A$, then we get a topological base of $\mathcal{A}$ as an $\mathcal{O}_1$-module, and this base satisfies the conditions (7.5.1) and (7.5.2). Thus we have to show that we can choose g_i in such a way that $g_i \in \Delta(\nu(g_i))$.

Recall that, according to (7.1.1), we have a standard base $\{f_1, f_2, \ldots, f_n\}$ of $\mathcal{A}$ such that $f_j \in \Delta(\nu(f_j))$. Let

$$\overline{f}_j = [f_j \bmod \mathcal{M}^{\nu(f_j)+1}]$$

be the initial form of f_j.

It is easy to see that we can choose $\overline{g}_i$ as

$$\overline{g}_i = \sum_{j=1}^{n} \overline{h}_{ij} \overline{f}_j$$

where $\overline{h}_{ij}$ is a homogeneous polynomial in $\overline{x}_1, \overline{x}_2, \ldots, \overline{x}_n$ with coefficients in $\mathbf{k}$ of degree $\deg \overline{h}_{ij} = \deg \overline{g}_i - \deg \overline{f}_j$.

Now by (7.1.4) we can find

$$h_{ij} \in \Delta(\deg \overline{h}_{ij})$$

whose initial form is $\overline{h}_{ij}$:

$$\overline{h}_{ij} = \left[h_{ij} \bmod \mathcal{M}^{\deg \overline{h}_{ij}+1}\right] .$$

Then we can take

$$g_i = \sum_{j=1}^{n} h_{ij} f_j$$

and this is what we need, as clearly

$$\overline{g}_i = \left[g_i \bmod \mathcal{M}^{\nu(g_i)+1}\right]$$

and

$$g_i \in \Delta(\nu(g_i))$$

as $f_j \in \Delta(\nu(f_j))$ and $h_{ij} \in \Delta(\deg \overline{g}_i - \deg \overline{f}_j) = \Delta(\nu(g_i) - \nu(f_j))$. ■

(7.7) Proof of Proposition 7.4: Let N be the same as in (7.5). By (4.6)

$$k_*(\mathcal{D} + \mathcal{O}; \mathcal{A}) \geq N + 1 \tag{7.7.1}$$

as the rank of the pairing between $\epsilon(\mathcal{D} + \mathcal{O})$ and $\mathrm{Init}_{\leq d} \mathcal{A}$ is N .

As we have to show that $k_*(\mathcal{D} + \mathcal{O}; \mathcal{A}) = N + 1$, we consider the ideal

$$\mathrm{FT}^{N+1}(\mathcal{D} + \mathcal{O}; \mathcal{A}) .$$

Let $g_1, g_2, \ldots$ be as in Lemma 7.5, and let $D_1, D_2, \ldots, D_N \in \mathcal{D}$ be such that $\epsilon(D_1), \epsilon(D_2), \ldots, \epsilon(D_N)$ form a basis of $\epsilon(\mathcal{D})$. Then $D_1, D_2, \ldots, D_N$ form a basis of $\mathcal{D}$, and $D_0 = 1, D_1, D_2, \ldots, D_N$ form a basis of $\mathcal{D} + \mathcal{O}$.

By Proposition 4.5 the ideal $\mathrm{FT}^{N+1}(\mathcal{D} + \mathcal{O}; \mathcal{A})$ is generated by all $(N + 1) \times (N + 1)$ determinants $\det(D_i g_{j_s})$ where $0 \leq i \leq N$, $1 \leq s \leq N + 1$, $1 \leq j_1 < j_2 < \ldots < j_{N+1}$.

Note that $g_j \in \Delta(\nu(g_j))$ and $D_i \in \mathcal{D} \subset \mathrm{Diff}^d_{\mathcal{O}/\mathcal{O}_1}$, and $\mathrm{Diff}^d_{\mathcal{O}/\mathcal{O}_1} \subset \mathrm{Diff}^d_{I_{\mathrm{tr}}(\Delta)}$ as all the variables which are transverse to Δ lie in $\mathcal{O}_1$ by (7.1.2) and (7.1.4). Thus each $D_i \in \mathrm{Diff}^d_{I_{\mathrm{tr}}(\Delta)}$, and contactness of Δ yields

$$D_i g_j \in \Delta(\nu(g_j) - \mathrm{ord}\, D_i)$$

and consequently

$$\det(D_i g_{j_s}) \in \Delta\Big(\sum_{s=1}^{N+1} \nu(g_{j_s}) - \sum_{i=0}^{N} \mathrm{ord}\, D_i\Big) . \tag{7.7.2}$$

Clearly,

$$\sum_{s=1}^{N+1} \nu(g_{j_s}) \geq \sum_{j=1}^{N+1} \nu(g_j) = \sum_{k=1}^{d} k \cdot \dim\left[(\mathrm{Init}_k \mathcal{A}) \bmod (V_1)\right] + d + 1$$

and

$$\sum_{i=0}^{N} \operatorname{ord} D_i = \sum_{k=1}^{d} k \cdot \dim[\epsilon(\mathcal{D})]_k$$

where $[\epsilon(\mathcal{D})]_k = \epsilon(\mathcal{D}) \cap (\operatorname{gr}_k \mathcal{O})^*$. However, $\epsilon(\mathcal{D})$ and $(\operatorname{Init}_{\leq d} \mathcal{A}) \bmod (V_1)$ are dual, and the pairing preserves the grading; thus

$$\sum_{k=1}^{d} k \cdot \dim[(\operatorname{Init}_k \mathcal{A}) \bmod (V_1)] = \sum_{k=1}^{d} k \cdot \dim[\epsilon(\mathcal{D})]_k$$

and

$$\sum_{s=1}^{N+1} \nu(g_{j_s}) - \sum_{i=0}^{N} \operatorname{ord} D_i \geq d+1 .$$

Now (7.7.2) yields

$$\det(D_i g_{j_s}) \in \Delta(d+1) .$$

Thus

$$\mathrm{FT}^{N+1}(\mathcal{D}+\mathcal{O};\mathcal{A}) \subset \Delta(d+1) .$$

In particular, this means that $\mathrm{FT}^{N+1}(\mathcal{D}+\mathcal{O};\mathcal{A}) \neq (1)$, so $k_*(\mathcal{D}+\mathcal{O};\mathcal{A}) \leq N+1$. Together with (7.7.1) this yields

$$k_*(\mathcal{D}+\mathcal{O};\mathcal{A}) = N+1$$

and

$$\mathrm{FT}^*(\mathcal{D}+\mathcal{O};\mathcal{A}) = \mathrm{FT}^{N+1}(\mathcal{D}+\mathcal{O};\mathcal{A}) \subset \Delta(d+1).$$

■

(7.8) Remark. Our next point is to define the filtration Δ_0 as the minimal contact filtration satisfying

$$\mathrm{FT}^*(\mathcal{D}+\mathcal{O};\mathcal{A}) \subset \Delta_0(d+1)$$

for any $\mathcal{D}$ and d satisfying the conditions of Proposition 7.4. However, the conditions on $\mathcal{D}$ included $\mathcal{D} \subset \mathrm{Diff}^d_{\mathcal{O}/\mathcal{O}_1}$ and the conditions (7.1.1)–(7.1.5) that we imposed on the subring $\mathcal{O}_1$, involved lots of additional data: a standard base $f_1, f_2, \ldots, f_n$, a filtration Δ and elements $x_1, x_2, \ldots, x_m$. Thus we first have to reformulate the conditions on $\mathcal{O}_1$ without reference to these additional data, and here is the answer.

(7.9) Remark. Let the subring $\mathcal{O}_1$ satisfy the conditions (7.1.1)–(7.1.5). It is not hard to see that it satisfies the following conditions:

(7.9.1) $\mathcal{O}_1$ is a normal crossing ring and for some $x_1, x_2, \ldots, x_m$, we have $\mathcal{O} = \mathcal{O}_1[[x_1, x_2, \ldots, x_m]]$ as normal crossing rings.

(7.9.2) $\mathcal{O}_1$ is in generic position to $\mathcal{A}$.

(7.9.3) Let $\mathcal{M}_1$ be the maximal ideal of $\mathcal{O}_1$, and let $W \subset \mathcal{M}/\mathcal{M}^2$ be as in (7.1); then

$$(\mathcal{M}_1 \bmod \mathcal{M}^2) + W = \mathcal{M}/\mathcal{M}^2 .$$

Conversely, it is not hard to see that if $\mathcal{O}_1$ satisfies these conditions, then we can find a standard base $\{f_1, f_2, \ldots, f_n\}$ of $\mathcal{A}$, a filtration Δ and elements $x_1, x_2, \ldots, x_n \in \mathcal{M}$ so that the conditions (7.1.1)–(7.1.5) are satisfied.

Now we can give the definition of the filtration Δ_0 .

7.10 Definition. *We define the filtration* Δ_0 *as the minimal contact filtration satisfying*

$$\mathrm{FT}^*(\mathcal{D} + \mathcal{O}; \mathcal{A}) \subset \Delta(d+1)$$

for any positive integer d *and any submodule* $\mathcal{D}$ *satisfying the following conditions:*

(7.10.1) $\mathcal{D} \subset \mathrm{Diff}^d_{\mathcal{O}/\mathcal{O}_1}$ *for some subring* $\mathcal{O}_1 \subset \mathcal{O}$ *satisfying conditions (7.9.1)–(7.9.3).*

(7.10.2) $\mathcal{D}$ *is d-dual to* $\mathcal{A}$.

(Such minimal filtration exists and is unique by the results of [Youssin 1], Section 9.)

7.11 Corollary to Proposition 7.4. *Any contact filtration* Δ *satisfying condition (2.7.1), contains this filtration* Δ_0 . ■

Thus this filtration Δ_0 is a good candidate to be the minimal filtration of the Main Theorem 2.7. In Section 9 we shall show that our filtration Δ_0 also satisfies condition (2.7.1), and this will prove the Main Theorem 2.7.

7.12 Remark. The constructions of this section are to some extent parallel to [Giraud 1], § 3. The subring $\mathcal{O}_1 \subset \mathcal{O}$ corresponds to the morphism $f : X \to S$ of [Giraud 1]; as we already mentioned in (6.6), the condition that this morphism is transverse, corresponds to the property of $\mathcal{O}_1$ being in generic position to $\mathcal{A}$. As we explained in (5.1), taking the Fitting ideals $\mathrm{FT}^*(\mathrm{Diff}^\ell_{\mathcal{O}}; \mathcal{A})$ corresponds to taking the absolute Hilbert-Samuel stratum of $X = \operatorname{Spec} \mathcal{O}/\mathcal{A}$; it is easy to see that the Fitting ideals $\mathrm{FT}^*(\mathrm{Diff}_{\mathcal{O}/\mathcal{O}_1}; \mathcal{A})$ correspond to the *relative* Hilbert-Samuel stratum of X/S (see [Giraud 1], Definition 2.1).

Now the difference of our approach with that of [Giraud 1] is as follows. The ideals $\mathrm{FT}^*(\mathrm{Diff}^\ell_{\mathcal{O}/\mathcal{O}_1}; \mathcal{A})$ contain the higher elements of a standard base differentiated many times (cf. (5.5)). More precisely, each f_i comes differentiated $\nu(f_i) - 1$ times. This is good for maximal contact, but this is not enough for our purposes as we want each f_i *itself* to be contained in $\Delta(\nu(f_i))$. This is the reason why we considered the Fitting ideals $\mathrm{FT}^*(\mathcal{D}; \mathcal{A})$ where $\mathcal{D}$ is a *submodule* in $\mathrm{Diff}^\ell_{\mathcal{O}/\mathcal{O}_1}$.

8. Normalized standard bases.

(8.1) In this section we more or less follow [Hironaka 1], Chapter III, § 7, introducing some slight generalizations. We shall assume we are given a complete regular local ring $\mathcal{O}$, a subring $\mathcal{O}_1 \subset \mathcal{O}$, $\mathcal{O} = \mathcal{O}_1[[x_1, x_2, \ldots, x_m]]$, and an ideal $\mathcal{A} \subset \mathcal{O}$ such that $\mathcal{O}_1$ and $\mathcal{A}$ are in generic position. Let $\mathcal{M}$ and $\mathcal{M}_1$ be the maximal ideals of $\mathcal{O}$ and $\mathcal{O}_1$ respectively, and let $\mathbf{k} = \mathcal{O}/\mathcal{M} = \mathcal{O}_1/\mathcal{M}_1$. Let

$$V = \mathcal{M}/\mathcal{M}^2 = \mathrm{gr}_1\, \mathcal{O}$$

$$V_1 = \mathcal{M}_1/\mathcal{M}_1^2 \subset \mathcal{M}/\mathcal{M}^2 = V$$

and let (V_1) be the ideal in $\mathrm{gr}\, \mathcal{O}$ generated by V_1 ; then, as in (7.2),

$$\frac{\mathrm{gr}\, \mathcal{O}}{(V_1)} = \mathbf{k}[\widehat{x}_1, \widehat{x}_2, \ldots, \widehat{x}_m] \tag{8.1.1}$$

where $\widehat{x}_i$ is the image of x_i in $(\mathrm{gr}\, \mathcal{O})/(V_1)$.

In general, for any $f \in \mathcal{O}$ we shall denote by $\overline{f}$ its initial form

$$\overline{f} = [f \bmod \mathcal{M}^{\nu(f)+1}]$$

and by $\widehat{f}$ its image in $(\operatorname{gr}\mathcal{O})/(V_1)$:

$$\widehat{f}=\left[\overline{f} \bmod (V_1)\right]\in\frac{\operatorname{gr}\mathcal{O}}{(V_1)}\,.$$

By (8.1.1) $\widehat{f}$ is a polynomial in $\widehat{x}_1,\widehat{x}_2,\ldots,\widehat{x}_m$:

$$\widehat{f}=\sum_{\alpha=(\alpha_1,\alpha_2,\ldots,\alpha_m)} c_\alpha \widehat{x}_1^{\alpha_1}\widehat{x}_2^{\alpha_2}\ldots\widehat{x}_m^{\alpha_m} \tag{8.1.2}$$

where $c_\alpha\in\mathbf{k}$.

8.2 Definition. *If $\widehat{f}\neq 0$, then we shall say that the lexicographically highest term in (8.1.2) is the principal term of f with respect to the presentation $\mathcal{O}=\mathcal{O}_1[[x_1,x_2,\ldots,x_m]]$. In other words, $x^\alpha=x_1^{\alpha_1}x_2^{\alpha_2}\ldots x_m^{\alpha_m}$ is the principal term of f if α is the lexicographically maximal such that $c_\alpha\neq 0$. We shall also say that in this case $\overline{x}^\alpha$ is the principal term of $\overline{f}\in\operatorname{gr}\mathcal{O}$.*

8.3 Remark. Of course, not every $f\in\mathcal{O}$ or $\overline{f}\in\operatorname{gr}\mathcal{O}$ has a principal term as $\widehat{f}$ may be equal to zero.

8.4 Definition. *We say that a standard base $\{f_1,f_2,\ldots,f_n\}$ of $\mathcal{A}$ is normalized with respect to $\mathcal{O}=\mathcal{O}_1[[x_1,x_2,\ldots,x_m]]$ if each f_i has the property that its power series expansion*

$$f_i=\sum_{\alpha=(\alpha_1,\alpha_2,\ldots,\alpha_m)} f_{i\alpha}x_1^{\alpha_1}x_2^{\alpha_2}\ldots x_m^{\alpha_m}$$

(here $f_{i\alpha}\in\mathcal{O}_1$) does not contain the principal term (if it exists) of any element of the ideal

$$(\overline{f}_1,\overline{f}_2,\ldots,\overline{f}_{i-1})\in\operatorname{gr}\mathcal{O}\,.$$

8.5 Proposition. *A normalized standard base exists for any presentation $\mathcal{O}=\mathcal{O}_1[[x_1,x_2,\ldots,x_m]]$ as above.*

8.6 Proof: Indeed, we start with any standard base $\{f_1,f_2,\ldots,f_n\}$. If f_i is the first element such that its power series expansion contains the principal term of some element of the ideal $(\overline{f}_1,\overline{f}_2,\ldots,\overline{f}_{i-1})$, then we can subtract from f_i a linear combination of $f_1,f_2,\ldots,f_{i-1}$ to kill any such term in its power series expansion. Although such operation may bring into the expansion of f_i new terms which are principal terms of elements of the ideal $(\overline{f}_1,\overline{f}_2,\ldots,\overline{f}_{i-1})$, these new terms will either have higher degree or they will be lexicographically lower than the one we killed. Thus a convergent infinite procedure will kill all such terms in the expansion of f_i , and by a similar procedure we can make sure that each f_j , $j=i+1,i+2,\ldots,n$ is free from the principal terms of the elements of the ideal $(\overline{f}_1,\overline{f}_2,\ldots,\overline{f}_{j-1})$. ■

8.7 Remark. It is not hard to see that a normalized standard base of a given ideal $\mathcal{A}$ with respect to a given presentation $\mathcal{O}=\mathcal{O}_1[[x_1,x_2,\ldots,x_m]]$ still does not have to be unique, as we can add to f_i a linear combination of $f_{i+1},f_{i+2},\ldots,f_n$.

9. Proof of the Main Theorem 2.7.

Let $\mathcal{O}$ and $\mathcal{A}$ be as above, let $\mathcal{O}_1$ be a subring of $\mathcal{O}$ satisfying the conditions (7.9.1)–(7.9.3) and suppose $\mathcal{O} = \mathcal{O}_1[[x_1, x_2, \ldots, x_m]]$ as normal crossing rings. Let $\{f_1, f_2, \ldots, f_n\}$ be a standard base of $\mathcal{A}$, normalized with respect to the presentation $\mathcal{O} = \mathcal{O}_1[[x_1, x_2, \ldots, x_m]]$, and let Δ_0 be as in (7.10).

9.1 Proposition. *$f_i \in \Delta_0(\nu(f_i))$ for all $i = 1, 2, \ldots, n$.*

9.2 Proof: Take any element f_i of the normalized standard base, and let $d = \nu(f_i) - 1$. Let

$$\widehat{x}^{\alpha^1}, \widehat{x}^{\alpha^2}, \ldots, \widehat{x}^{\alpha^N}$$

(here $\alpha^j = (\alpha_1^j, \alpha_2^j, \ldots, \alpha_m^j)$) be the principal terms of all elements of

$$\left[(\mathrm{Init}_{\leq d}\mathcal{A}) \bmod (V_1)\right] \subset \mathbf{k}[\widehat{x}_1, \widehat{x}_2, \ldots, \widehat{x}_m] .$$

It is a standard result in the linear algebra that

$$N = \dim\left[(\mathrm{Init}_{\leq d}\mathcal{A}) \bmod (V_1)\right] .$$

Moreover, there exists a homogeneous basis $\{\widehat{g}_1, \widehat{g}_2, \ldots, \widehat{g}_N\}$ of the linear space

$$\left[(\mathrm{Init}_{\leq d}\mathcal{A}) \bmod (V_1)\right] \subset \mathbf{k}[\widehat{x}_1, \widehat{x}_2, \ldots, \widehat{x}_m]$$

such that the principal term of $\widehat{g}_j$, $1 \leq j \leq N$, is

$$\widehat{x}^{\alpha^j} = \widehat{x}_1^{\alpha_1^j}\widehat{x}_2^{\alpha_2^j}\ldots\widehat{x}_m^{\alpha_m^j} .$$

Let $g_1, g_2, \ldots, g_N \in \mathcal{O}$ be the lifts of $\widehat{g}_1, \widehat{g}_2, \ldots, \widehat{g}_N$; we may choose $g_1, g_2, \ldots, g_N \in \mathcal{A}$.

Let D_j , $j = 1, 2, \ldots, N$, be defined by

$$D_j = \left(\frac{\partial}{\partial x}\right)^{\alpha^j} = \left(\frac{\partial}{\partial x_1}\right)^{\alpha_1^j}\left(\frac{\partial}{\partial x_2}\right)^{\alpha_2^j}\cdots\left(\frac{\partial}{\partial x_N}\right)^{\alpha_N^j} \in \mathrm{Diff}^d_{\mathcal{O}/\mathcal{O}_1}$$

and let

$$\mathcal{D} = \mathcal{O}\cdot D_1 + \mathcal{O}\cdot D_2 + \ldots + \mathcal{O}\cdot D_N .$$

Then clearly $\mathcal{D}$ is d-dual to $\mathcal{A}$. Let $D_0 = 1$ and consider the determinant

$$h = \det\begin{pmatrix} D_1g_1 & D_1g_2 & \ldots & D_1g_N & D_1f_i \\ D_2g_1 & D_2g_2 & \ldots & D_2g_N & D_2f_i \\ \ldots & \ldots & \ldots & \ldots & \ldots \\ D_Ng_1 & D_Ng_2 & \ldots & D_Ng_N & D_Nf_i \\ D_0g_1 & D_0g_2 & \ldots & D_0g_N & D_0f_i \end{pmatrix} \tag{9.2.1}$$

$$\in \mathrm{FT}^{N+1}(\mathcal{D}+\mathcal{O};\mathcal{A}) = \mathrm{FT}^*(\mathcal{D}+\mathcal{O};\mathcal{A}) .$$

We claim that the expansion of f_i in the powers of $x_1, x_2, \ldots, x_m$ does not contain any term divisible by any of x^{α^j} , $j = 1, 2, \ldots, N$; this is because $f_1, f_2, \ldots, f_n$ form a normalized standard base, so f_i does not contain any term which is a principal term of any element of the ideal $(\overline{f}_1, \overline{f}_2, \ldots, \overline{f}_{i-1}) \subset \mathrm{gr}\,\mathcal{O}$, which shows that f_i does not contain any term divisible by the principal term of any element of $(\mathrm{Init}_{\leq d}\mathcal{A}) \bmod (V_1)$, i.e., f_i does not contain any term divisible by x^{α^j}, $j = 1, 2, \ldots, N$.

This shows that

$$D_1 f_i = D_2 f_i = \ldots = D_N f_i = 0$$

and our determinant can be simplified:

$$h = h' \cdot D_0 f_i = h' \cdot f_i$$

where

$$h' = \det \begin{pmatrix} D_1 g_1 & D_1 g_2 & \ldots & D_1 g_N \\ D_2 g_1 & D_2 g_2 & \ldots & D_2 g_N \\ \ldots & \ldots & \ldots & \ldots \\ D_N g_1 & D_N g_2 & \ldots & D_N g_N \end{pmatrix}.$$

This determinant is invertible; to see this, we reduce all its elements modulo $\mathcal{M}$. Then it becomes block-triangular (blocks correspond to those of $\overline{g}_j$ and D_j which have the same degree), and the determinants of the diagonal blocks are nonzero elements of the field $\mathbf{k} = \mathcal{O}/\mathcal{M}$. This means that the whole determinant h' is nonzero modulo $\mathcal{M}$ and thus invertible (cf. (7.7)). Finally, by (9.2.1) and (7.10)

$$f_i = (h')^{-1} h \in \mathrm{FT}^*(\mathcal{D} + \mathcal{O}; \mathcal{A}) \subset \Delta_0(d+1) = \Delta_0(\nu(f_i)) .$$

■

9.3 Proof of the Main Theorem 2.7: Indeed, Δ_0 is a contact filtration and by (9.1) satisfies the condition (2.7.1); by (7.11) it is contained in any other filtration satisfying these properties. ■

Appendix: Sketch of another proof.

Here we present another proof of the Main Theorem 2.7 which does not use Fitting ideals, but rather some refinement of the concept of the normalized standard base.

A.1 Notation. We shall assume that $\mathcal{O}$ is a complete normal crossing ring, $\mathcal{A}$ is an ideal in $\mathcal{O}$ and $\mathcal{O}_1 \subset \mathcal{O}$ is a subring satisfying conditions (7.9.1)–(7.9.3). Let $x_1, x_2, \ldots, x_m$ be such elements of $\mathcal{O}$ that $\mathcal{O} = \mathcal{O}_1[[x_1, x_2, \ldots, x_m]]$ as normal crossing rings. Let $\mathcal{M}$ and $\mathcal{M}_1$ be the maximal ideals of $\mathcal{O}$ and $\mathcal{O}_1$ respectively.

Both the following definition and Proposition A.3 are due to [Bierstone, Milman 1].

A.2 Definition. *We say that a standard base $\{g_1, g_2, \ldots, g_n\}$ is completely normalized with respect to the presentation $\mathcal{O} = \mathcal{O}_1[[x_1, x_2, \ldots, x_n]]$, if the following property is satisfied. Let*

$$x^{\alpha^i} = x_1^{\alpha_1^i} x_2^{\alpha_2^i} \ldots x_m^{\alpha_m^i}$$

be the principal term of g_i (see Definition 8.1). Then we require that the power series expansion of g_i has the form

$$g_i = x^{\alpha^i} + \sum_{\alpha=(\alpha_1,\alpha_2,\ldots,\alpha_m)} g_{i\alpha} x_1^{\alpha_1} x_2^{\alpha_2} \ldots x_m^{\alpha_m} = x^{\alpha^i} + \sum g_{i\alpha} x^\alpha$$

where the second term $\sum g_{i\alpha} x^\alpha$ does not involve any x^α which is a principal term of any element of $\mathrm{Init}\,\mathcal{A}$.

Note that what we call a *completely normalized standard base*, [Bierstone, Milman 1] call a *standard base*.

A.3 Proposition. *The ideal $\mathcal{A}$ has a unique standard base which is completely normalized with respect to the presentation $\mathcal{O} = \mathcal{O}_1[[x_1, x_2, \ldots, x_m]]$.*

A.4 Proof: We can choose a base $\{\overline{g}_1, \overline{g}_2, \ldots, \overline{g}_n\}$ of the ideal $\operatorname{Init}\mathcal{A}$ in such a way that the principal term of each $\overline{g}_i$ is x^{α^i} and the principal term of any element of $\operatorname{Init}\mathcal{A}$ lies in the ideal

$$(x^{\alpha^1}, x^{\alpha^2}, \ldots, x^{\alpha^n}) . \tag{A.4.1}$$

Let $\Gamma \subset \mathbf{Z}_+^m$ be the set of the exponents of all monomials that do not lie in the ideal (A.4.1). (Here by $\mathbf{Z}_+$ we denote the set of nonnegative integers.) Then the generalized Weierstrass preparation theorem of [Hironaka 4], Section 4, states that the map

$$\sum_{\alpha\in\Gamma} \mathcal{O}_1 \cdot x^\alpha \to \mathcal{O}/\mathcal{A} \tag{A.4.2}$$

is an isomorphism of $\mathcal{O}_1$-modules.

Let

$$\sum_{\alpha\in\Gamma} g_{i\alpha} x^\alpha$$

be such an element whose image under the map (A.4.2) is equal to

$$-x^{\alpha^i} \bmod \mathcal{A}$$

and let

$$g_i = x^{\alpha^i} + \sum g_{i\alpha} x^\alpha .$$

Then clearly $g_i \in \mathcal{A}$ and it is easy to see that $g_1, g_2, \ldots, g_n$ form a completely normalized standard base of $\mathcal{A}$.

Conversely, if $\{g_1, g_2, \ldots, g_n\}$ is a completely normalized standard base, then it is easy to see that the principal terms of $g_1, g_2, \ldots, g_n$ are $x^{\alpha^1}, x^{\alpha^2}, \ldots, x^{\alpha^n}$ (possibly up to a permutation). Clearly, then

$$g_i = x^{\alpha^i} + \sum_{\alpha\in\Gamma} g_{i\alpha} x^\alpha$$

where $\sum g_{i\alpha} x^\alpha$ is the preimage of $-x^{\alpha^i} \bmod \mathcal{A}$ under the map (A.4.2). Thus the completely normalized standard base is unique. ∎

Now we come to the key point of our proof.

A.5 Proposition. *Let Δ be a contact filtration in $\mathcal{O}$ satisfying the property (2.7.1) (i.e., such that for some standard base $\{f_1, f_2, \ldots, f_n\}$ of $\mathcal{A}$ we have $f_i \in \Delta(\nu(f_i))$). Let $\{g_1, g_2, \ldots, g_n\}$ be the completely normalized standard base of $\mathcal{A}$ with respect to the presentation $\mathcal{O} = \mathcal{O}_1[[x_1, x_2, \ldots, x_m]]$. Then $g_i \in \Delta(\nu(g_i))$.*

The idea of the proof of this proposition is to approximate the completely normalized standard base $\{g_1, g_2, \ldots, g_n\}$ by other standard bases, say, $\{f_1', f_2', \ldots, f_n'\}$, which satisfy $f_i' \in \Delta(\nu(f_i'))$.

A.6 Definition. *We shall say that a standard base $\{f_1', f_2', \ldots, f_n'\}$ of $\mathcal{A}$ is completely normalized up to degree s, if we can find such elements $f_{i\alpha}' \in \mathcal{O}_1$ that*

$$f_i' \equiv x^{\alpha^i} + \sum f_{i\alpha}' x^\alpha \pmod{\mathcal{M}^s}$$

and $\sum f'_{i\alpha}x^\alpha$ does not involve any x^α which is a principal term of any element of $\operatorname{Init}\mathcal{A}$.

A.7 Lemma. *Suppose $\{f'_1, f'_2, \dots, f'_n\}$ is a standard base of $\mathcal{A}$ which is completely normalized up to degree $s \geq 1$, and suppose Δ is a contact filtration such that $f'_i \in \Delta(\nu(f'_i))$. Then there exists a standard base $\{f''_1, f''_2, \dots, f''_n\}$ which is completely normalized up to degree $s+1$ and satisfies*

$$f''_i \in \Delta(\nu(f''_i)) \tag{A.7.1}$$

and

$$f''_i \equiv f'_i \pmod{\mathcal{M}^s} . \tag{A.7.2}$$

A.8 Proof: Indeed, take such $f'_{i\alpha} \in \mathcal{O}_1$ that

$$f'_i \equiv x^{\alpha^i} + \sum f'_{i\alpha}x^\alpha \pmod{\mathcal{M}^s}$$

and such that $\sum f'_{i\alpha}x^\alpha$ does not involve any x^α which is a principal term of any element of $\operatorname{Init}\mathcal{A}$; by multiplying f'_i by an invertible element of $\mathcal{O}$, we may assume that the expansion of f'_i involves x^{α^i} with coefficient 1 exactly. We may also assume that $\sum f'_{i\alpha}x^\alpha \in \mathcal{M}^{\nu(f'_i)}$.

Denote

$$\overline{h}_i = \left[(f'_i - x^{\alpha^i} - \sum f'_{i\alpha}x^\alpha) \bmod \mathcal{M}^{s+1}\right] . \tag{A.8.1}$$

Then $\overline{h}_i \in \operatorname{gr}_s \mathcal{O}$. Consider the expansion of $\overline{h}_i$ in the powers of $\overline{x}_i = [x_i \bmod \mathcal{M}^2]$. Suppose this expansion involves any monomials which are principal terms of elements of $\operatorname{Init}\mathcal{A}$, and let $\overline{x}^\beta$ be the lexicographically highest among them. Then $\beta \neq \alpha^i$ and $\deg x^\beta \geq \deg x^{\alpha^i}$; if $\deg x^\beta = \deg x^{\alpha^i}$ then β is lexicographically lower than α^i . Now $\overline{x}^\beta$ is the principal term of some element of $\operatorname{Init}\mathcal{A}$, and — as one can easily see from the condition (7.9.3) — of some element of $\operatorname{Init}\mathcal{A} \cap \mathbf{k}[W]$. Thus $\overline{x}^\beta$ is the principal term of

$$\sum \overline{a}_j \overline{f}'_j$$

where $\overline{a}_j$ is a homogeneous element of $\mathbf{k}[W]$ of degree $s - \nu_j$, where $\nu_j = \nu(f'_j)$.

Note that

$$W \subset [\Delta(1) \bmod \mathcal{M}^2]$$

as Δ contains some standard base; thus any homogeneous element of $\mathbf{k}[W]$ of degree, say, ν , can be lifted to an element of $\Delta(\nu)$.

Thus we can lift each $\overline{a}_j$ to some element $a_j \in \Delta(s - \nu_j)$, and we may assume $a_j = 0$ if $\nu_j > s$. Let

$$f'''_i = f'_i - \sum a_j f'_j .$$

Then it is easy to see that $\{f'''_i \ , \ i = 1, 2, \dots, n\}$ is a standard base of $\mathcal{A}$, which is completely normalized up to degree s , $f'''_i \in \Delta(\nu_i)$, and

$$f'''_i \equiv x^{\alpha^j} + \sum f'''_{i\alpha}x^\alpha \bmod \mathcal{M}^s$$

where $\sum f'''_{i\alpha}x^\alpha$ does not involve principal terms of elements of $\operatorname{Init}\mathcal{A}$. Moreover, the expansion of

$$\left[(f'''_i - x^{\alpha^i} - \sum f'''_{i\alpha}x^\alpha) \bmod \mathcal{M}^{s+1}\right] \in \operatorname{gr}_s \mathcal{O}$$

(cf. (A.8.1)) involves only such principal terms of elements of $\operatorname{Init}\mathcal{A}$ that are lexicographically smaller than x^{β}.

Thus we see that applying this procedure again and again to the standard base $\{f_i''', i = 1,2,\dots,n\}$, we shall finally get the standard base $\{f_i'', i = 1,2,\dots,n\}$, which has all the same properties, and, in addition,

$$\left[(f_i'' - x^{\alpha^i} - \sum f_{i\alpha}'' x^{\alpha}) \bmod \mathcal{M}^{s+1}\right]$$

is free from the principal terms of the elements of $\operatorname{Init}\mathcal{A}$. This, however, means that the base $\{f_i'', i = 1,2,\dots,n\}$, is normalized up to degree $s+1$. Finally, it is not hard to see that $f_i'' \equiv f_i' \bmod \mathcal{M}^s$. ■

A.9 Proof of Proposition A.5: Indeed, take a standard base $\{f_1, f_2, \dots, f_n\}$ such that $f_i \in \Delta(\nu(f_i))$. As a standard base of $\mathcal{A}$, it is completely normalized up to degree 1. Applying Lemma A.7 to this base, we get inductively a sequence of standard bases which are completely normalized up to degrees $2,3,4,\dots$ respectively. Moreover, according to (A.7.2), this sequence is convergent. Clearly, the limit is a completely normalized standard base, say, $\{f_1^{(\infty)}, f_2^{(\infty)}, \dots, f_n^{(\infty)}\}$; by (A.7.1) it satisfies

$$f_i^{(\infty)} \in \Delta(\nu(f_i^{(\infty)}) .$$

Finally, the uniqueness of the completely normalized standard base yields $g_i = f_i^{(\infty)}$, thus

$$g_i \in \Delta(\nu(g_i)) .$$

■

A.10 Proof of the Main Theorem 2.7: We are given an ideal $\mathcal{A}$ in a complete normal crossing ring $\mathcal{O}$ with the maximal ideal $\mathcal{M}$. As above, let W be the minimal subspace in $\mathcal{M}/\mathcal{M}^2$ which is weakly transverse to the fixed variables and such that $\operatorname{Init}\mathcal{A}$ can be generated by polynomials in the elements of W.

Choose a presentation $\mathcal{O} = \mathcal{O}_1[[x_1, x_2, \dots, x_m]]$ as a normal crossing ring (then $\mathcal{O}_1$ is also a normal crossing ring), such that the images of $x_1, x_2, \dots, x_m$ in $\mathcal{M}/\mathcal{M}^2$ form a basis of the subspace $W \subset \mathcal{M}/\mathcal{M}^2$. Then clearly $\mathcal{M}_1/\mathcal{M}_1^2 \subset \mathcal{M}/\mathcal{M}^2$ is a complementary subspace to W. (Here, as above, $\mathcal{M}_1$ is the maximal ideal of $\mathcal{O}_1$). By (6.7) $\mathcal{O}_1$ is in generic position to $\mathcal{A}$; moreover, $\mathcal{O}_1$ clearly satisfies conditions (7.9.1)–(7.9.3).

By Proposition A.3 there exists a completely normalized standard base $\{g_1, g_2, \dots, g_n\}$ of $\mathcal{A}$ (with respect to the presentation $\mathcal{O} = \mathcal{O}_1[[x_1, x_2, \dots, x_m]]$). Let Δ_0 be the minimal contact filtration satisfying

$$g_i \in \Delta_0(\nu(g_i)) .$$

Clearly, Δ_0 satisfies the condition (2.7.1); we claim that Δ_0 is the minimal contact filtration satisfying this condition.

Indeed, let Δ be any contact filtration satisfying (2.7.1); then there exists a standard base $\{f_1, f_2, \dots, f_n\}$ of $\mathcal{A}$ satisfying

$$f_i \in \Delta(\nu(f_i)) .$$

Then by Proposition A.5

$$g_i \in \Delta(\nu(g_i)) .$$

Thus $\Delta_0 \subset \Delta$, which means that Δ_0 is minimal. ■

References

[Bennett] B. M. Bennett, *On the characteristic function of a local ring,* Ann. of Math. **91** (1970), pp. 25–87.

[Bierstone, Milman 1] E. Bierstone, P. D. Milman, *Relations among analytic functions I,* Ann. Inst. Fourier, **37** (1987), pp. 187–239.

[Bierstone, Milman 2] E. Bierstone, P. D. Milman, *Uniformization of analytic spaces,* Journal of the Amer. Math. Soc. (to appear).

[Bierstone, Milman 3] E. Bierstone, P. D. Milman, in preparation.

[EGA IV_{IV}] A. Grothendieck, J. Dieudonné, *Elements de géométrie algébrique* IV_{IV}, Publ. Math. IHES **32** (1967).

[Fitting] M. Fitting, *Die Determinantenideale eines Moduls,* Jahresbericht Deutsch. Math.-Verein. **46** (1936), pp. 195–228.

[Giraud 1] J. Giraud, *Sur la theorie du contact maximal,* Math Zeit. **137** (1974), pp. 285–310.

[Giraud 2] J. Giraud, *Etude local des singularités,* Publ. Math. d'Orsay, 1971/72, réimpression 1980.

[Giraud 3] J. Giraud, *Contact maximal en caractéristique positive*, Ann. Scient. Ec. Norm. Sup., 4e Série, **8** (1975), pp.201–234.

[Hironaka 1] H. Hironaka, *Resolution of singularities of an algebraic variety over a field of characteristic zero,* Ann. of Math. **79** (1964), pp. 109–326.

[Hironaka 2] H. Hironaka, *Bimeromorphic smoothing of a complex-analytic space,* Acta Math. Vietnamica **2** (1977), pp. 103–168.

[Hironaka 3] H. Hironaka, *Certain numerical characters of singularities,* J. Math. Kyoto Univ. **10** (1970), 151–187.

[Hironaka 4] H. Hironaka, *Idealistic exponents of singularity,* Algebraic geometry, The Johns Hopkins Centennial Lectures, Johns Hopkins University Press, 1977, pp. 52–125.

[Singh] B. Singh, *Effect of a permissible blowing-up on the local Hilbert functions,* Inventiones Math. **26** (1974), pp. 201–212.

[Villamayor 1] O. Villamayor Jr., *On resolution of singularities,* unpublished manuscript, 1981.

[Villamayor 2] O. Villamayor Jr., *On resolution of singularities,* Inst. Argentino de Matematica, 1987, Preprint No. 110.

[Youssin 1] B. Youssin, *Newton polyhedra without coordinates,* in this volume.

[Youssin 2] B. Youssin, *Canonical formal uniformization in characteristic zero,* in preparation.

Boris Youssin
Courant Institute of
Mathematical Sciences
New York University
251 Mercer Street
New York, NY 10012

Current address (1989/90) :
Institute of Mathematics
Hebrew University
Givat-Ram
Jerusalem, Israel

MEMOIRS of the American Mathematical Society

SUBMISSION. This journal is designed particularly for long research papers (and groups of cognate papers) in pure and applied mathematics. The papers, in general, are longer than those in the TRANSACTIONS of the American Mathematical Society, with which it shares an editorial committee. Mathematical papers intended for publication in the Memoirs should be addressed to one of the editors:

General instructions to authors for

PREPARING REPRODUCTION COPY FOR MEMOIRS

> **For more detailed instructions send for AMS booklet, "A Guide for Authors of Memoirs." Write to Editorial Offices, American Mathematical Society, P.O. Box 6248, Providence, R.I. 02940.**

MEMOIRS are printed by photo-offset from camera copy fully prepared by the author. This means that, except for a reduction in size of 20 to 30%, the finished book will look exactly like the copy submitted. Thus the author will want to use a good quality typewriter with a new, medium-inked black ribbon, and submit clean copy on the appropriate model paper.

Model Paper, provided at no cost by the AMS, is paper marked with blue lines that confine the copy to the appropriate size. Author should specify, when ordering, whether typewriter to be used has **PICA**-size (10 characters to the inch) or **ELITE**-size type (12 characters to the inch).

Line Spacing — For best appearance, and economy, a typewriter equipped with a half-space ratchet — 12 notches to the inch — should be used. (This may be purchased and attached at small cost.) Three notches make the desired spacing, which is equivalent to 1-1/2 ordinary single spaces. Where copy has a great many subscripts and superscripts, however, double spacing should be used.

Special Characters may be filled in carefully freehand, using dense black ink, or **INSTANT** ("rub-on") **LETTERING** may be used. AMS has a sheet of several hundred most-used symbols and letters which may be purchased for $5.

Diagrams may be drawn in black ink either directly on the model sheet, or on a separate sheet and pasted with rubber cement into spaces left for them in the text. Ballpoint pen is not acceptable.

Page Headings (Running Heads) should be centered, in CAPITAL LETTERS (preferably), at the top of the page — just above the blue line and touching it.

LEFT-hand, EVEN-numbered pages should be headed with the AUTHOR'S NAME;

RIGHT-hand, ODD-numbered pages should be headed with the TITLE of the paper (in shortened form if necessary).

Exceptions: PAGE 1 and any other page that carries a display title require NO RUNNING HEADS.

Page Numbers should be at the top of the page, on the same line with the running heads.

LEFT-hand, EVEN numbers — flush with left margin;

RIGHT-hand, ODD numbers — flush with right margin.

Exceptions: PAGE 1 and any other page that carries a display title should have page number, centered below the text, on blue line provided.

FRONT MATTER PAGES should be numbered with Roman numerals (lower case), positioned below text in same manner as described above.

MEMOIRS FORMAT

> **It is suggested that the material be arranged in pages as indicated below. Note: Starred items (*) are requirements of publication.**

Front Matter (first pages in book, preceding main body of text).

Page i — *Title, *Author's name.

Page iii — Table of contents.

Page iv — *Abstract (at least 1 sentence and at most 300 words).

Key words and phrases, if desired. (A list which covers the content of the paper adequately enough to be useful for an information retrieval system.)

**1980 Mathematics Subject Classification (1985 Revision)*. This classification represents the primary and secondary subjects of the paper, and the scheme can be found in Annual Subject Indexes of MATHEMATICAL REVIEWS beginnning in 1984.

Page 1 — Preface, introduction, or any other matter not belonging in body of text.

Footnotes: *Received by the editor date.
Support information — grants, credits, etc.

First Page Following Introduction - Chapter Title (dropped 1 inch from top line, and centered). Beginning of Text.

Last Page (at bottom) - Author's affiliation.